Advanced Well Completion Engineering

Advanced Well Completion Engineering

Editor

Shivraj Choudhary

Advanced Well Completion Engineering

Edited by **Shivraj Choudhary**

Printed in 2017

ISBN: 978-1-68117-331-3

Library of Congress Control Number: 2015939243

Contents

Preface..vii

Chapter 1   A Combined Probabilistic and Optimization Approach for
             Improved Chemical Mixing Systems Design.......................................1
             Matthew J. Opgenorth, William E. McDermott, Peter Laz, and
             Corinne S. Lengsfeld

Chapter 2   Integrative Approach to the Plant Commissioning Process..............23
             Kris Lawry and Dirk John Pons

Chapter 3   Chemical Analysis on Mongolia's Natural Bitumen 49
             Erdenetsogt Bat-Erdene, Batdelger Byambagar, Erdenee Enkhtsetseg,
             and Budeebazar Avid

Chapter 4   Hydrodynamic Cavitation-Assisted Synthesis of Nanocalcite..........59
             Shirish H. Sonawane, Sarang P. Gumfekar, Kunal H. Kate, Satish P.
             Meshram, Kshitij J. Kunte, Laxminarayan Ramjee1, Candrashekhar
             M. Mahajan, Madan G. Parande, and Muthupandian Ashokkumar

Chapter 5   Influence of the Chemical Composition of Completion Fluids on
             the Propagation of Electromagnetic Waves within Oil Wells...........79
             Alexandre Ashade Lassance Cunha, Marco Aurélio Pacheco, and
             José Ricardo Bergmann

Chapter 6   Effects of Chemical Reaction on the Unsteady Free Convection
             Flow past an Infinite Vertical Permeable Moving Plate with
             Variable Temperature 95
             Fayza Mohammed Nasser El-Fayez

Chapter 7   Microwave Plasma Enhanced Chemical Vapor Deposition of
             Carbon Nanotubes..115
             Ivaylo Hinkov, Samir Farhat, Cristian P. Lungu, Alix Gicquel,
             François Silva, Amine Mesbahi, Ovidiu Brinza, Cornel Porosnicu,
             and Alexandru Anghel

Chapter 8   **Studies on Chemical Resistance of PET-Mortar Composites: Microstructure and Phase Composition Changes**..........................143

Yassine Senhadji, Ilies Bahlouli, and David Houivet

Chapter 9   **Synthesis and Characterization of GaN Rods Prepared by Ammono-Chemical Vapor Deposition**..191

Gregorio Guadalupe Carbajal Arízaga, Karina Viridiana Chávez Hernández, Nicolás Cayetano Castro, Manuel Herrera Zaldivar, Rafael García Gutiérrez, and Oscar Edel Contreras López

Chapter 10   **Selection Method of Surfactants for Chemical Enhanced Oil Recovery**..217

Roland Nagy, Rubina Sallai, László Bartha, and Árpád Vágó

Citations..231

Index..235

Preface

Well completion engineering is an important component part of oil and gas well construction and a basis of field development implementation. It has a goal of ensuring regular and safe production and prolonging the production life of oil and gas wells. The traditional mode of well completion engineering, which had been adopted in China for a long time, cannot meet the requirements of developing circumstances; thus, reform is needed. After summing up domes-tic and foreign experience and lessons, the new Advanced Well Completion concept has been presented. Based on field geology and reservoir engineering, it adopts the nodal analysis method, and drilling, well completion, and production are organically integrated, thus forming an integrated well completion engineering system.

Editor

A Combined Probabilistic and Optimization Approach for Improved Chemical Mixing Systems Design

Matthew J. Opgenorth[1], William E. McDermott[2], Peter Laz[1], and Corinne S. Lengsfeld[1]

[1]Department of Mechanical and Materials Engineering, University of Denver, Denver, USA

[2]Applied Research and Technology Institute, University of Denver, Denver, USA

ABSTRACT

A design analysis of a mixing nozzle was performed using a combination of probabilistic and optimization techniques. A novel approach was utilized where probabilistic analysis was used to reduce the number of geometric constraints based on sensitivity factors. An optimization algorithm used only the most significant parameters to maximize mixing. A second probabilistic analysis was performed after optimization was complete in order to quantitatively predict the effects of manufacturing tolerances on mixing performance. This process for automated design is attractive over full parameter optimization techniques due to the computational efficiency resulting from an intelligent reduction in evaluated variables.

INTRODUCTION

Computational fluid dynamics (CFD) analyses, with software packages like Fluent™ (ANSYS, Inc.), have become accepted techniques for solving complex fluid flows using numerical implementation of fluid mechanics principles. CFD-driven optimization has been explored by several researchers [1-3]. Peigin et al. has proposed a design tool that implements multi-constrained optimization of shape design driven by Genetic Algorithms (GA) coupled with CFD. The benefit to GA is that they can handle a large number (20+) of design variables; however, with a large number of constraints, GA are computationally expensive. Peigin et al. reports 15-18 hours for a single point optimization and on average 8-12 optimization steps [1]. A similar multi-constrained optimization strategy has been implemented on a ship's hull. Again the downside is the time constraints, where they report approximately 10 days for 20 shape generations on a PC-based cluster [2]. Many researchers are using these GA's in order to optimize a system with numerous variables.

GA's have also been used by Carroll (1996) for the modeling of complex chemical mixing systems. The GA was coupled with the Blaze II code [4]. The Blaze II code can address up to 500 chemical reactions and 40 species. It also contains 1-D fluid dynamic equations, with mixing terms derived from 2-D equations, that can be used to model axis symmetric and 2-D flow fields [4]. The GA used 5 parameters with either 32 or 16 discrete possibilities resulting in approximately 2 million permutations. Even with a 2-D model and only 5 variable parameters the GA required 8 days of continuous runtime [3].

The objective of the current project is to develop and implement a design analysis of a fluid mixing nozzle using a coupled optimization and probabilistic approach. Due to the potentially large computational times to run GA optimization, our approach is to reduce the number of constraints though probabilistic analysis in order to use single objective optimization techniques. This technique was applied to mixing nozzles for a generic chemical mixing system. Based on the insight into the features that most affect performance provided by the probabilistic analysis, optimization has the potential to improve design performance while also reducing the weight of the system. Additionally, a probabilistic analysis of the optimized design can confirm the original optimization parameters and provide insight into the effects of manufacturing tolerances of specific geometric variables. This will in turn, reduce manufacturing costs by determining which dimensions are important to the mixing performance of the nozzle and which geometric tolerances can be loosened. A clear benefit to this design approach is that it allows for fast optimization by reducing the number of constraints through probabilistic analysis, as well as, assessing the impact of manufacturing tolerances.

METHODOLOGY

Currently, there is no commercially available tool in industry to perform fluid mechanics analyses, optimization and probabilistic analysis in the proposed integrated manner. There are however, software packages that can execute individual components of the process. Our approach is to use commercially available CFD, probabilistic and optimization software and interface them together with custom scripting. The probabilistic software, or optimization routine, manages the CFD code by importing any number of variables within ranges and/ or distributions set by the user. Figure 1 and 2 show a diagram linking the processes mentioned above.

CFD Model

The CFD software used for the simulations was Fluent™ Version 6.3.26 [5]. It is a widely used computational software package for modeling fluid flow and heat transfer in complex geometries. Easy mesh generation and ability to refine or coarsen the mesh autonomously based on the flow solution are just some of the features that make this CFD package extremely versatile and ideal for automation. Gambit ® was the pre-processor used for the solid modeling and mesh generation.

The algorithm employed was the pressure-based Navier-Stokes solution algorithm. Typically, this algorithm is used for low velocity incompressible flows, which fits this case where the flow velocity is about 83 m/s. The momentum equation provides the velocity field and the density is calculated from the equation of state. Other governing equations include energy and species conservation. The turbulence model chosen was the standard k-epsilon turbulence model. This model is robust and suitable for initial iterations, initial alternative design screenings, and parametric studies. The k-epsilon model will be ideal for the automated analysis where many different shapes will be analyzed. A no-slip boundary condition was placed at the walls. The primary inlet utilized a user-defined function (UDF) that modeled a fully developed fluid flow.

Definition of the boundary conditions is critical to both accuracy of the model and computational efficiency, the importance of which should not be under estimated when venturing into optimization and probabilistic analysis. Considerable attention was paid to identifying the best algorithmic parameters in order to achieve convergence in the minimum number of iterations, reliably, and with physically correct solutions regardless of geometry. It is also noteworthy to mention that this process was carried out in parallel on 8 processors in order to reduce computational time.

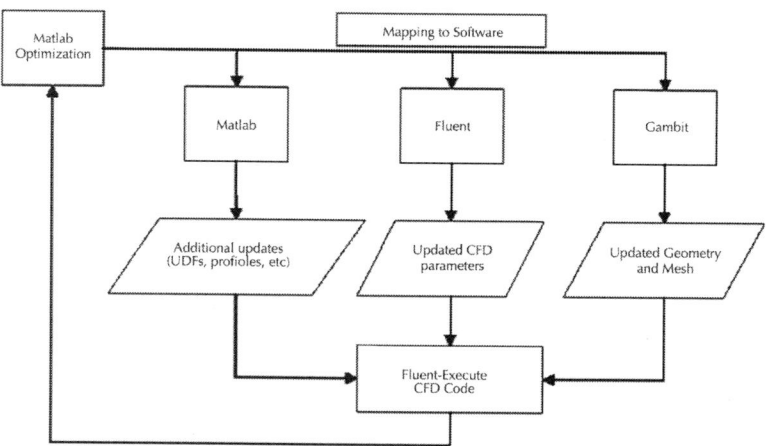

Figure 1: Diagram of the probabilistic/optimization and CFD interface. Each program is linked by custom scripts.

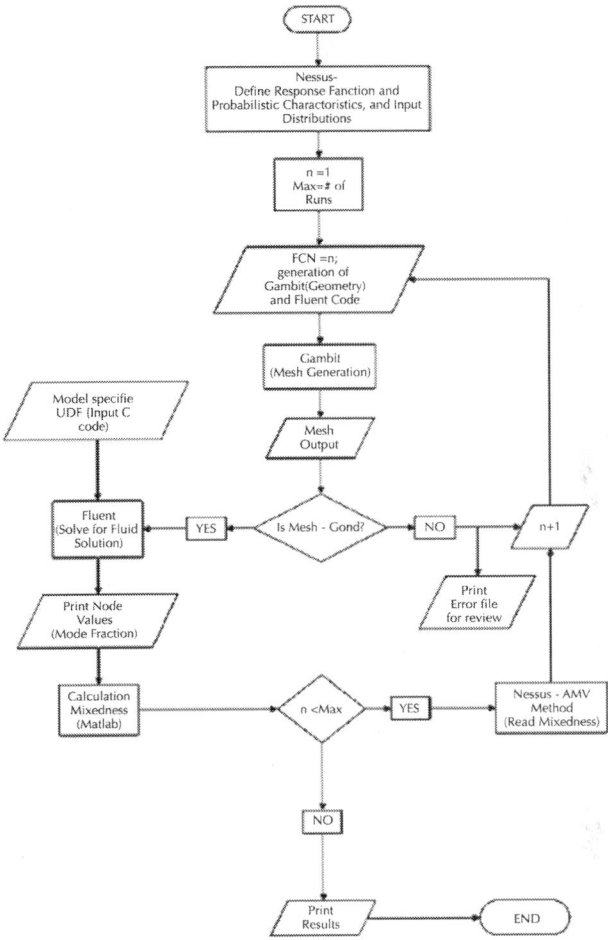

Figure 2: Example flow chart of the probabilistic interface using the AMV method.

Probabilistic Model

As uncertainty is inherent in physical systems, a probabilistic analysis model input variables as distributions and then predicts a distribution of performance. Based on the distribution of performance, sensitivity and importance factors are used to identify critical parameters. The probabilistic software, Nessus®, implements a variety of probabilistic methods that vary in efficiency and accuracy of the solution. The most commonly used probabilistic method is Monte Carlo [6]. The Monte Carlo method generates random values for each variable according to its distribution and then predicts the distribution of performance through repeated trials. As the accuracy of the prediction is dependent on the number of trials performed, the Monte Carlo method is computationally expensive. The mean-value (MV) family of methods are approximate, but considerably more efficient than the Monte Carlo method. They create a mean-based response function and compute the most probable point (MPP), which is the shortest distance from the origin to the limit state surface and represents the combination of stochastic variables resulting in a specific level of performance [6]. In this study, the Advanced Mean Value (AMV) method was applied, which uses a higher order approximation to determine the MPP [7]. The number of trials used for the AMV method is 1 + the number of variables + the number of probability levels desired [7]. While the AMV method is a discrete and approximate solution, it has shown excellent agreement with Monte Carlo analysis for monotonic system. This has been shown in other applications which utilize FE analysis with geometric perturbations and realistic loading conditions [8, 9]. The major advantage of the AMV is that it requires a small number of trials, which saves significant computational time when the CFD analysis requires multiple iterations to reach convergence. Both of these probabilistic methods provide sensitivity factors identifying and quantifying the contributions of each variable on performance of the system.

The AMV method was utilized early in the design process to identify the variables that contributed significantly to the behavior of the system reducing the number of parameters required for performance optimization. By reducing the number of variables in the optimization, the computation time and required resources are dramatically reduced. Instead of using a sensitivity study perturbing a single variable at a time, a probabilistic approach was used to allow for the interaction affects between the various input parameters. During the early stages of the design process, the emphasis was on identifying the important and unimportant variables. Following optimization, AMV methods were employed to evaluate the effects of manufacturing tolerances on system performance.

Optimization Model

Numerical optimization techniques are designed to minimize an objective function subject to constraints, with many algorithms developed over the past several decades [10]. In general, the algorithms require a starting point, x_0, and then iterate or step until there is no more progression, or the approximate solution falls within a user-defined tolerance. Typically, algorithms follow one of two types of strategies, line search or trust region. This study implemented a trust region [11] strategy in order to account for geometric changes that may result in the fluid domain acting non-linearly. A common problem in line searches is the fixed step size can causes them to miss a local minimum, whereas the step size in the trust region search is not fixed and, therefore, has a better opportunity to find a minimum that is close to the current point.

The success and efficiency of an optimization is contingent on selection of an appropriate algorithm and an accuracy characterization of the problem. It was not known whether the variables would behave linear, nonlinear, or convex, only that they could hold any value between the bounds. The optimization algorithm had to be suitable for a continuous objective function with variables that are constrained by simple bounds and can solve for linear, nonlinear and convex variables.

A trust-region algorithm was implemented, which utilizes an active-set algorithm, for the optimization analysis. An active-set algorithm will employ linear techniques to estimate the active-set at each iteration and then solve an equality constrained quadratic program to generate a step [12]. This method was used because it tends to yield more exact solutions and is less sensitive to the initial starting point than interior point methods. Another benefit, in our case, of the active-set algorithm is that it uses a gradient projection method when only bounds are applied to the constraints [13]. The gradient projection method attempts to speed up the solution process within the active-set, but is only utilized when the variables are bounded. It consists of two different stages. First, the search direction will be along the path of steepest decent from the current point. The second stage investigates the face of the feasible region using the active-set constraints [12]. This can significantly reduce the optimization time.

Interfacing Model

To facilitate communication between the all of the software packages, custom interfacing was developed to build CFD models with perturbed parameters and calculate performance parameters from the analysis outputs. Interfacing was performed with components written in Matlab®, Dos and C. In addition, checks were performed to ensure mesh quality to prevent analyses that would fail or highly skewed elements, which may lead to convergence issues. This is noteworthy because the automated process can potentially take days and even weeks to run and computational efficiency will be a driving factor, especially as more complex flows are examined. An imported UDF specifies the physical boundary conditions for the specific system. Once the CFD simulation converges to a solution for the flow field; it calculates the values of a user defined fluid property (i.e. mole fraction of interested species) via a UDF and prints the results to a file. A script is also utilized to calculate the performance parameter and print the results for analysis by either the probabilistic or optimization routines. The performance parameter used is Mixedness, which is defined by:

$$Mixedness = 1 - \frac{\sum \left| M_f - M_{f\_Homogeneous} \right|}{n \cdot M_{f\_Homogeneous}} \tag{1}$$

The degree of mixing is measured by the ratio of the integral value for species mole fraction (M_f) across an exit plane divided by the homogeneous mole fraction ($M_{f\_Homogeneous}$), where n is the number of nodes within the exit plane. Increased mixing of species in chemical systems should result in greater chemical efficiencies and better performance. This interfacing routine is continued until all of the probabilistic or optimization trials are completed.

PROBLEM DESCRIPTION

CFD Model

The combined probabilistic and optimization approach is demonstrated for a low flow, subsonic Hydrogen-Iodide Chlorine (HICl) Laser. As opposed to other chemical lasers the HICl Laser has yet to reach its expected performance potentially due in part to a lack of homogeneous distribution of its excited chemical species.

Additionally:

- The subsonic HICl Laser has simple mixing geometry, employing a repetitive nozzle array that allows the computational domain to be reduced based on simple lines of symmetry;
- The cross-flow injection geometry would be easy to perturb within the code;
- Validation data exists for injection cross-flow at speeds similar to the HICl.

The geometry for the subsonic HICl Laser consists of a rectangular flow channel (Figure 3(a)). There are four rows of secondary flow injection nozzles on the top and bottom plates of the cavity (Figure 3(b)). The primary inlet flow for the subsonic HICl Laser is a mixture of helium (He) and hydrogen (H_2). The secondary flow through the nozzle plates has two different mixtures that are being injected into the primary flow. The first three rows inject a mixture of helium and hydrogen-iodide (HI), where the fourth injects a mixture of helium and nitrogen-tri chloride (NCl_3). Table 1shows the boundary conditions applied during this design optimization run. A constant cross-sectional area was maintained on each inlet, but the aspect ratio was allowed to vary.

It would be computationally expensive to model the entire flow channel for the CFD analysis; consequently, planes of symmetry are used again to reduce the size of the modeled fluid domain (Figure 4). The fluid domain was modeled and meshed using text commands within a journal file. This allowed for easy modification during the automated process.

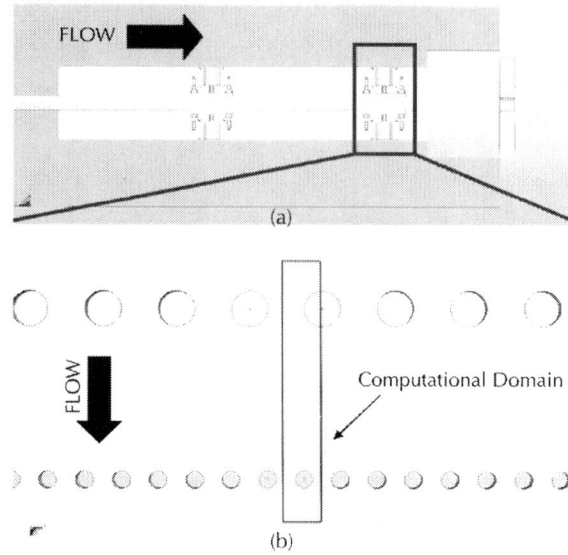

Figure 3: (a) A section view of the laser cavity showing the secondary inlet nozzle plates, and (b) detailed view of one of the nozzle plates. Note: the system has two identical sets of nozzle plates.

Table 1: Flow conditions for subsonic HICl laser

Initial Conditions and Constants		
Pressure	50	torr
Outlet Velocity	83	m/s
Temperature	300	K
Gas Constant	62400	Torr-cm^3/K-mole
Avogadro's Number	6.02E+23	molecules/mole

The algorithms, boundary conditions, and meshing strategies for the CFD simulation were validated using experimental data from subsonic (yet compressible flows) jets in cross flow [14]. Downstream velocity and temperature profiles obtained from the CFD simulations showed good comparison (Figure 5) to the experimental data of Dizene et al. (2000). This effort highlighted the need to accurately describe the velocity profile at the entrance boundaries and confirmed that the dynamic adaptive grid technique and algorithms selected worked well in these types of fluid flow conditions.

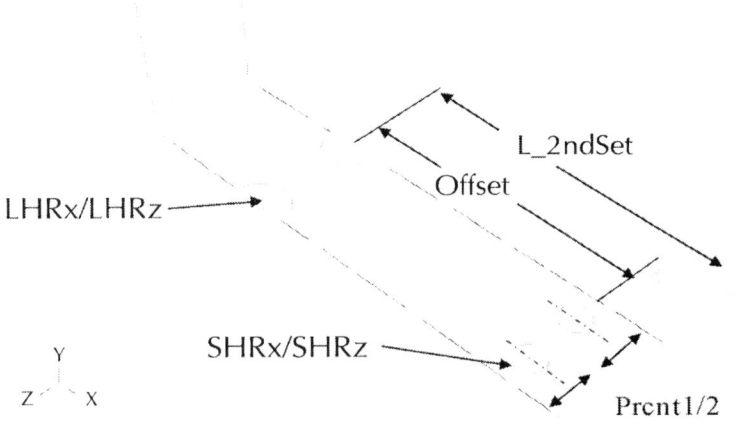

Figure 4: Schematic of the variables and geometry for the first probabilistic analysis.

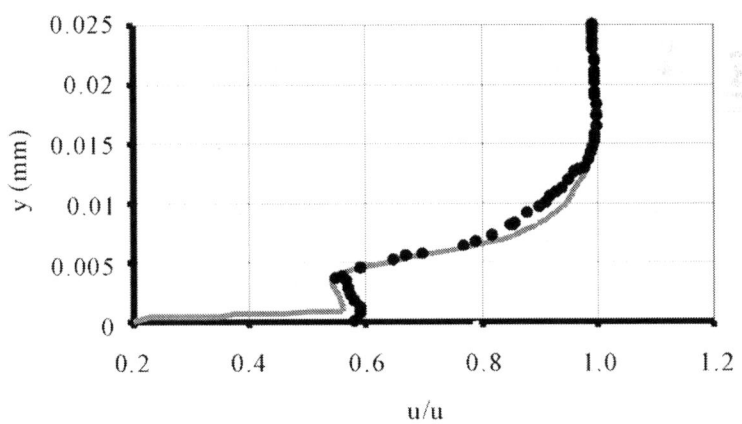

(a)

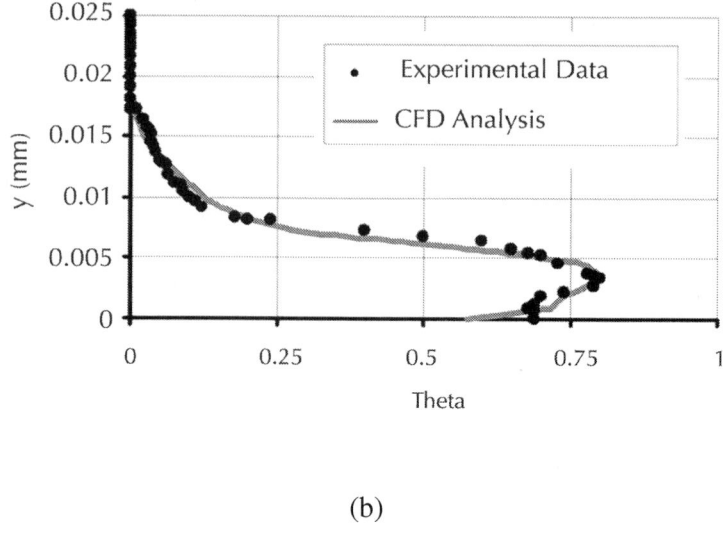

(b)

Figure 5: Velocity and Temperature profiles from experimental data compared to CFD simulations of jet in crossflow at X/D = 4.

Probabilistic Model

The initial probabilistic CFD analysis modeled eight geometric variables as distributions. The parameters were the size of the holes in two directions, the offset between the holes, and the hole placement, as illustrated in Figure 4. All of the parameters were modeled as normal distributions defined by mean and standard deviation (Table 2). Standard deviations were determined by the maximum range permissible by geometric limits.

Optimization Model

Optimization was then applied to determine the values of the variables within the constraints that maximized Mixedness. The four important variables, LHRx, LHRz, SHRx, and LHRz can be combined into two by keeping the cross sectional areas of the orifices constant. Therefore, only two variables, LHRz and SHRz, need to be perturbed for the HICl Laser optimization. The objective function and constraints for the optimization is stated below:

$$\text{Maximize: } M = f(\text{LHRz, SHRz}) \quad (2)$$

$$\text{Subject to: } 0.0055 \text{ cm} \leq \text{LHRz} \leq 0.023 \text{ cm}$$
$$0.010 \text{ cm} \leq \text{SHRz} \leq 0.042 \text{ cm} \quad (3)$$

The optimization routine was used to determine the values of LHRz and SHRz in each analysis. The initial guess in the optimization algorithm were the mean values from the probabilistic analysis. The automated process updated the geometry, generated the mesh, performed the CFD analysis and extracted the results to compute the Mixedness parameter. Since the maximum Mixedness is desired, the minimization of the ratio of HI concentrations to the maximum concentration (the second term in Equation (1) was calculated. The optimization process continues until the Mixedness value has converged to within a 0.01% tolerance. In addition, initial guesses near the lower and upper bounds of the design space were also evaluated to check for local minimums that may be on either side of the mean orifice dimensions.

Table 2: Probabilistic variables with respective mean, standard deviation, and distribution

Name	Mean (mm)	StDev (mm)	Distribution
LHRx	0.1125	0.067	Normal
LHRz	0.1125	0.067	Normal
SHRx	0.1	0.07	Normal
SHRz	0.1	0.07	Normal
OffsetH	0.2025	0.05	Normal
L_2ndSet	1.5	0.33	Normal
Prcnt1	0.0	0.3	Normal
Prcnt2	0.0	0.3	Normal

RESULTS AND DISCUSSION

Identification of Important Parameters

Figure 6 shows the importance levels associated with each variable that was perturbed in the modeled geometry. The importance factors represent the design vector or value of each parameter that defines the MPP, which is proportionate to the output measure at the specified probability. As they are reported in the standard normal space, importance factors are relative measures and the sum of the squares for each measure will equal 1. The importance levels identify the variables that contribute the most to the reliability of the design; therefore, it is deduced from Figure 5 that the radii of the inlet orifices contribute the most to either increase or decrease of the response variable, Mixedness. The other important observation is that PRCNT1, PRCNT2, OFFSETH and L_2NDSET had little to no effect on the Mixedness of the system. Recall, they had the largest standard deviation of all the variables. Now, the elimination of four of the eight variables within the geometry allows for the focus on only the radii of the orifices.

Design Optimization

All of the optimization analyses converged to the same value with the ratio of LHRz to SHRz equal to 0.6. This lower bound is both a computational and manufacturing limit due to the small size of the orifices. Figure 7 shows the contours of all the data taken from the three different optimization runs. The green area shows the larger Mixedness parameters and the path that the optimization algorithm followed. The optimization required between 18 and 29 analyses and from 5 to 9 optimization iterations.

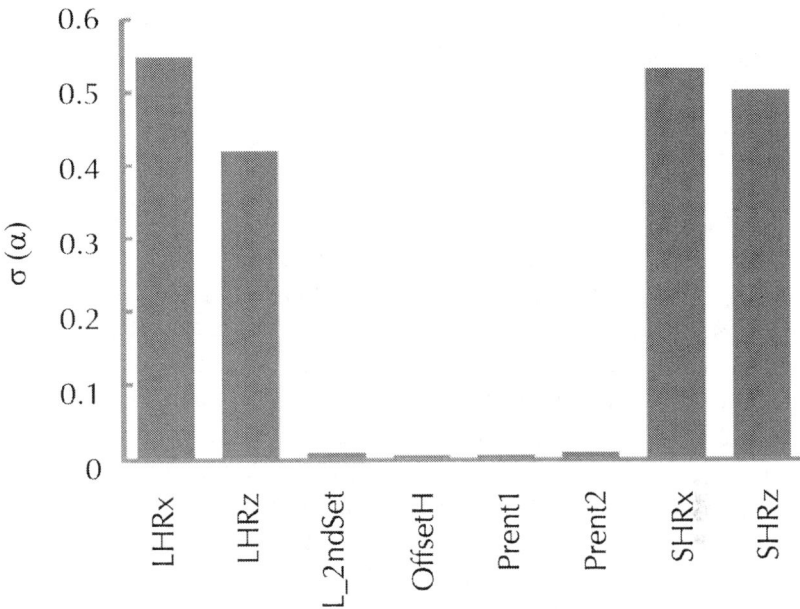

Figure 6: Histogram of the Importance Levels due to the first probabilistic analysis. It can be seen that the orifices affect the Mixedness the most.

The contour plot of the optimization analysis shows the optimized solution and the various starting points (Figure 8). Since, the only information acquired was along the optimization path, little reliable information can be extracted from the other areas of the plot. This does, however; show the optimal orifice aspect ratio for LHRz and SHRz is approximately equal to 4 and 6, respectively, and resulted in an increased Mixedness of nearly 10%. The major diameters of both orifices are parallel to the fluid flow.

In order to ensure that the global optimum was found, a Monte Carlo simulation was performed using the probabilistic CFD model to characterize the design space used in the optimization analysis. The response surface (Figure 9) with 200 Monte Carlo trials and the optimization results confirmed the findings of the optimization. Notably, the horizontal bands indicate that the LHRZ dimension has little effect on the fluid system. For example, starting at the optimized location (SHRz = 0.010 cm, LHRz = 0.006 cm), changes in LHRz within the design space did not affect the Mixedness value by more than 1%. Knowledge of these relationships is helpful in the design and manufacturing processes. The LHRz orifice can be a simple, circular geometry (which will be easier to create) and the mixing of the system will stay approximately the same compared to if it were elliptical. The response surface is non-monotonic over the optimization design space. The Monte Carlo results also reaffirm that the optimal configuration of the orifices, where the small orifice is located upstream to large orifice.

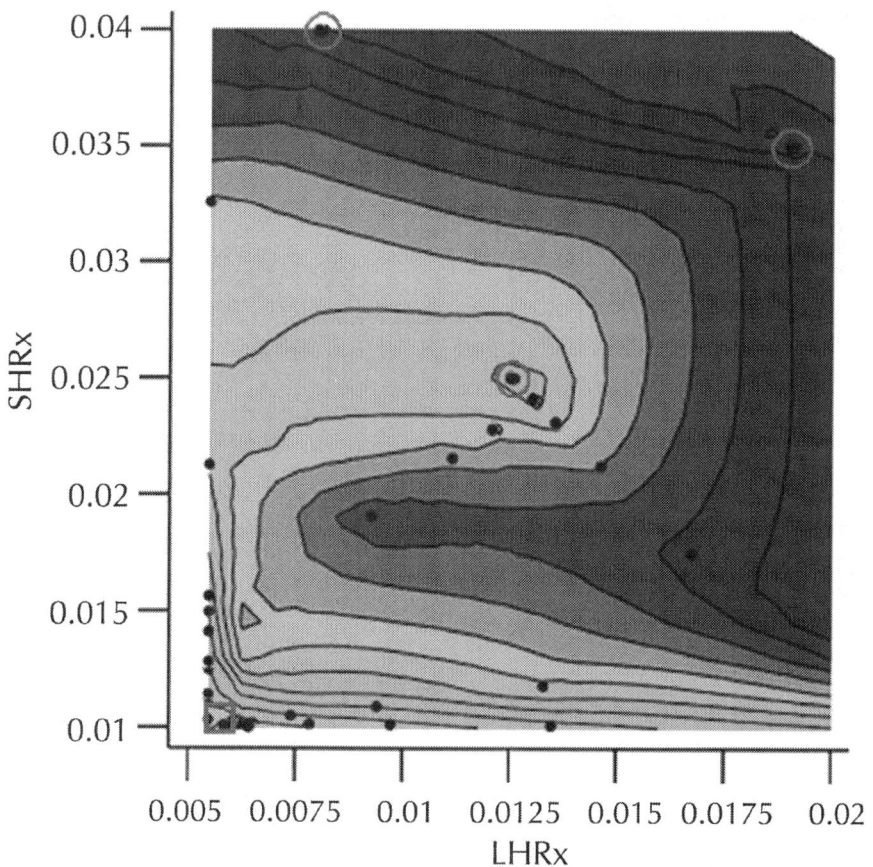

Figure 7: Contour plot of the optimization analysis with data point of each iteration. The optimized geometry converged near the lower bounds, where both orifices are elliptical and the major diameter is parallel to the fluid flow. The red circles locate the three different starting points and the red box indicates the optimum.

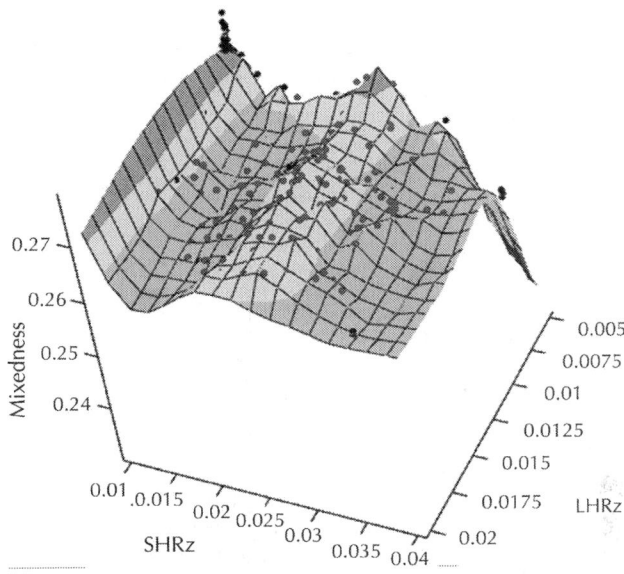

Figure 8: Surface plot of the combined Monte Carlo and optimization simulations with data points.

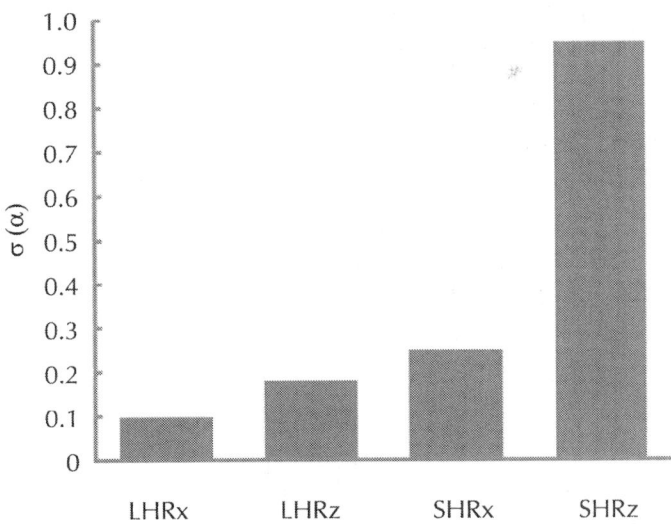

Figure 9: Histograms of the importance levels from the manufacturing sensitivity simulation.

Manufacturing Sensitivities

Based on the optimized design, the probabilistic platform was used to investigate the variability in Mixedness due to the manufacturing tolerances. The most economical way to create these elliptical geometries at the small dimensions required is through a wire electrical discharge machine (EDM). In this analysis, the geometry from the optimization simulation with mean values for LHRz and SHRz equal to 0.006 and 0.010 cm, respectively, will be used. The tolerances on these dimensions are +/− 0.001 cm, a tolerance that can typically be achieved by a wire EDM in a cost effective manner. The AMV method incorporated seven CFD simulations. The values for the AMV simulation are given below in Table 3.

The results for the importance of the four parameters, LHRz, LHRx, SHRz, and SHRz, are show in Figure 1. These results illustrate the same trends as the whole system. First, the LHRx, LHRz, and SHRx dimensions have virtually no importance on the Mixedness of the system. For the manufacturing design, the relaxation of tolerances on the LHRz dimensions would reduce the cost of fabrication and still maintaining a high Mixedness number. The importance of SHRz dimension does affect the system and the resulting Mixedness number. This result is liberal because the orifices' aspect ratios did change slightly. Since the X and Z dimensions were not dependent on each other, the X or Z radii could get larger or smaller while the other could do the same. Therefore, the results showed an increase in Mixedness, but if the aspect ratios were the same the Mixedness should not change. However, the Mixedness number reduces by ~3% if the SHRz dimension is decreased by 0.001 cm and increases by ~3% if it is decreased by 0.001 cm.

CONCLUSIONS

The purpose of this research was to create and implement an efficient computational design tool to combine optimization and probabilistic modeling to provide insight into how to improve chemical mixing systems performance. This has provided insight into the sensitivities of several different parameters that affect chemical mixing systems. Baseline CFD models were created for subsonic and mixing nozzles. The interfacing tools designed worked with minimal interaction from the user. The automated process for the probabilistic analysis requires input variables and perturbations from the user (assuming the geometry is created). Once all the inputs are specified, the interface carries out numerous evaluations of a fluid system. The optimization interface operated in much the same way. It required a starting point, as well as, constraints on the system. These actions were implemented on a subsonic mixing nozzle and results were obtained. Finally, the parallel processing technique enabled complex flows to be optimized in a timeframe consistent with real-time design processes (i.e. less than one week). For the AMV probabilistic analysis with 8 variables, the clock time was approximately 5 hours. The optimization process had a clock time on the order of 3-4 days. All the computations took place on a HP xw8400 Workstation with 2 - 2.33 GHz Xeon Quad Core Processors and 4 GB RAM.

Table 3: Probabilistic variables for manufacturing sensitivities on the optimized orifice geometry

Name	Mean	StDev	Distribution
LHRz	0.0060	0.00034	Normal
LHRx	0.0265	0.00034	Normal
SHRz	0.0636	0.00034	Normal
SHRx	0.0100	0.00034	Normal

It was discovered, using the above computational tools, that in a HICl laser mixing nozzle, elliptical orifices with the major diameter parallel to the flow direction increased the mixing within the system by roughly 10%. Haven and Kurosaka (1997) similarly found through experimentation that the injection port geometry had a powerful influence on penetration in the near field [15]. For this case the optimum aspect ratio of the larger orifice to be approximately 6 and showed that the small injection orifice should be placed in front of the larger in a staggered alignment pattern.

The second evaluation phase of this work explored the impact of manufacturing tolerances on Mixedness. It was shown that the tolerance on SHRz plays the largest role on mixing quality, and that the shape of the small diameter secondary injection orifice does not have a great effect on the Mixedness. In fact, the Mixedness only decreases by 0.7%.

ACKNOWLEDGEMENTS

Financial support for this project was funded by the DOD Joint Technology Office; "Hybrid Iodine Laser testing," Contract No. FA8632-05-C-2461.

REFERENCES

1.   S. Peigin and B. Epstein, "Multiconstrained Aerodynamic Design of Business Jet by Cfd Driven Optimization Tool," Aerospace Science and Technology, Vol. 12, No. 2, 2008, pp. 125-134. doi:10.1016/j.ast.2007.03.001

2.   Y. Tahara, D. Peri, E. Campana and F. Stern, "Computational Fluid Dynamics-Based Multiobjective Optimization of a Surface Combatant Using a Global Optimization Method," Journal of Marine Science and Technology, Vol. 13, No. 2, 2008, pp. 95-116.doi:10.1007/s00773-007-0264-7

3.   D. L. Carroll, "Chemical Laser Modeling with Genetic Algorithms," AIAA Journal, vol. 34, no. 2, 1996, pp. 338-346.

4.   L. H. Sentman, M. Subbiah and S. W. Zelazny, "Blaze II: A Chemical Laser Simulation Computer Program," Bell Aerospace Textron, TR H-CR-77-8, 1977.

5.   FLUENT_Inc., FLUENT Documentation, 2006.

6.   A. Haldar and S. Mahadevan, "Probability, Reliability and Statistical Methods in Engineering Design," John Wiley & Sons, Inc., New York, 2000.

7.   Y.-T. Wu, H. R. Millwater and T. A. Cruse, "Advanced Probabilistic Structural Analysis Method for Implicit Performance Functions," AIAA Journal, vol. 28, no. 9, 1990, pp. 1663-1669.

8.   S. K. Easley, et al., "Finite Element-Based Probabilistic Analysis Tool for Orthopaedic Applications," Computer Methods and Programs in Biomedicine, Vol. 85, No. 1, 2007, pp. 32-40. doi:10.1016/j.cmpb.2006.09.013

9.   P. J. Laz, J. Q. Stowe, M. A. Baldwin, A. J. Petrella and P. J. Rullkoetter, "Incorporating Uncertainty in Mechanical Properties for Finite Element-Based Evaluation of Bone Mechanics,"

Journal of Biomechanics, Vol. 40, No. 13, 2007, pp. 2831-2836. doi:10.1016/j.jbiomech.2007.03.013

10. J. Nocedal and S. J. Wright, "Numerical Optimization. Springer Series in Operations Research," Springer-Verlag, New York, 1999.

11. G. N. Vanderplaats, "Numerical Optimization Techniques for Engineers," Third Edition, Vanderplaats Research and Development, Inc., 2001.

12. R. H. Byrd and R. A. Waltz, "An Active-Set Algorith for Nonlinear Programming Using Parametric Linear Programming," Department of Industrial and Systems Engineering, University of Southern California, Los Angeles, 2007.

13. MATLAB, The MathWorks Inc., 1994-2008

14. R. Dizene, J. M. Charbonnier, E. Dorignac and R. Leblanc, "Experimental Study of Inclined Jets Cross Flow Interaction in Compressible Regime. I. Effect of Compressibility in Subsonic Regime on Velocity and Temperature Fields," International Journal of Thermal Sciences, Vol. 39, No. 3, 2000, pp. 390-403. doi:10.1016/S1290-0729(00)00219-2

15. B. A. Haven and M. Kurosaka, "Kidney and Anti-Kidney Vortices in Crossflow Jets," Journal of Fluid Mechanics, Vol. 352, 1997, pp. 27-64. doi:10.1017/S0022112097007271

Integrative Approach to the Plant Commissioning Process

Kris Lawry[1] and Dirk John Pons[2]

[1]Department of Chemical and Process Engineering, University of Canterbury, Private Bag 4800, Christchurch 8020, New Zealand

[2]Department of Mechanical Engineering, University of Canterbury, Private Bag 4800, Christchurch 8020, New Zealand

ABSTRACT

Commissioning is essential in plant-modification projects, yet tends to be ad hoc. The issue is not so much ignorance as lack of systematic approaches. This paper presents a structured model wherein commissioning is systematically integrated with risk management, project management, and production engineering. Three strategies for commissioning emerge, identified as direct, advanced, and parallel.

Direct commissioning is the traditional approach of stopping the plant to insert the new unit. Advanced commissioning is the commissioning of the new unit prior to installation. Parallel commissioning is the commissioning of the new unit in its operating position, while the old unit is still operational. Results are reported for two plant case studies, showing that advanced and parallel commissioning can significantly reduce risk. The model presents a novel and more structured way of thinking about commissioning, allowing for a more critical examination of how to approach a particular project.

INTRODUCTION

Background

Plant modifications are an ongoing process throughout the life of any process plant. Reasons for modification include efforts to improve reliability, production capacity, quality, or productivity. Seamless incorporation is the key concern associated with the installation of any new equipment in an operating plant due to the high cost of process downtime. Several steps can be taken to minimise the risk associated with the installation of new equipment such as hazard and operability studies, project management, development of redundancy plans, and commissioning of the new equipment.

Of these, commissioning is an essential activity in many plant-modification projects and has significant implications for project success. Yet paradoxically it tends to be approached in an ad hoc manner. It is often included in project plans, so it is not that people are ignorant of commissioning. Rather, the problem is that there is a lack of systematic approaches to commissioning, so it is frequently left to tradespeople and plant operators to manage in whatever way they see fit. This is an undesirable situation since it results in unpredictable outcomes. In some cases it can even cause serious problems. An extreme example would be the catastrophic failure of the Chernobyl nuclear power plant (1986), which was caused by operators attempting an ad hoc test of the efficacy of a modified emergency cooling system.

This paper presents a structured conceptual model for the commissioning process, and two cases studies showing application to operating plant.

EXISTING MODELS OF COMMISSIONING

Literature

Many authors have highlighted the value of commissioning from a range of different perspectives but they all agree that commissioning and the integration of a new project is critical to the success of any project [1–10].However commissioning is poorly defined and is interpreted ambiguously [6, 11], which leads to inefficient utilisation within industry. In this paper "commissioning" is defined as the disciplined activity involving careful testing, calibration, and proving of all systems, software, and networks within the project boundary [5].

Current Models of Commissioning

Factors that are known to affect the commissioning process include the following.

- Type of project. Thus situational variables are important; that is, the factors that resulted in a successful (or failed) commissioning outcome in one case are not necessarily transferrable to a different situation.

- Who is in charge of the phase. Commissioning can be completed by a range of different groups depending on the project. It can be the equipment manufacturers, operation team, or a separate commissioning team depending on scale and requirements of the project [12]. The relationships between these people are also important (social dimension) [6, 13], hence also contractual obligations (see (iv) below).

- Number/type of phases. Commissioning can also be broken down in several sections such as planning, precommissioning, testing, integration, monitoring, documenting, and handover depending on the level of complexity of the project. This requires careful project planning (see (iv) below).

- Project planning and contractual sufficiency. It is widely recognised in the literature that commissioning requires deliberate planning, as opposed to ad hoc treatment. Thus it needs appropriate consideration in the work breakdown structure and project planning [14], allocation of resources, transferral of those costs into the initial contract [9, 15–18], and creation of specific operating procedures (especially important for safety-critical plant like boilers [19]). This corresponds to the "integration" tasks in the project management approach [6].

The commissioning process has been examined for a wide range of different projects [2–5, 8, 20, 21]. The predominate approach can be described as task specific; the literature tends to identify specific tasks that should be completed as part of commissioning. Thus the focus has been on completing multiple checks on a system to ensure it will operate as expected. Thus there are many reports in the literature, too numerous to mention, about commissioning experiences in specific case studies. These are undoubtedly helpful, especially for lessons learned and application to comparable situations. They are also systematic, in a way, especially in the provision of templates and checklists to guide practitioners.

However there is a lack of holistic or integrative models. There is much less literature at the next higher level of abstraction, which is the commissioning process in general. At this level we are interested not so much in case-specific experiences but in the fundamental principles and the methodology. What exists at this level is primarily in the area of instrumentation and control; some examples are [5, 22, 23].

Thus the existing commissioning strategies in the literature can be categorised into three types, see Figure 1. These are (a) ad hoc, which is action-orientated problem solving; and (b) template, which involves using a checklist, or operating procedure that worked before or in another situation. Both (a) and (b) are premised on the assumption that commissioning is a routine set of tasks. The third strategy challenges that premise and calls for a deliberately thoughtful approach. Thus the third category in the literature is (c) methodological, which involves analysis of the situational needs and deliberate selection of the most relevant of several possible commissioning methods.

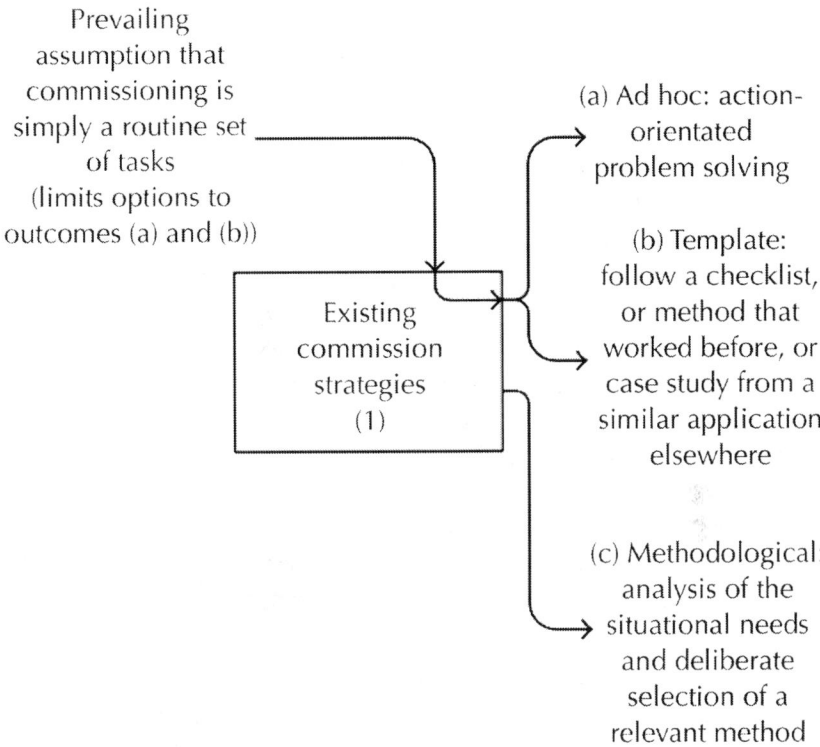

Figure 1: Three common strategies for commissioning, broadly reflecting the approaches described in the literature. If commissioning is perceived to be simply a routine set of tasks, which is a common assumption, then this tends to preclude any more thoughtful approach to the problem.

Issues and Problem Areas

A clear refrain in the literature is that commissioning (i) needs deliberate project management, but (ii) is too often not given the attention it deserves. One of the issues with commissioning, which contributes to problem (ii), is that the value thereof is hard to quantify. Justifying the value of commissioning may be completed using qualitative analysis similar to quantification of risk in a project [24]. This is based on the consequence and probability of the system failing to operate as anticipated. In other cases there is no attempt at justification at all, so the value is not appreciated.

Another issue is the tendency to underresource the commissioning in the project planning, which is issue (i) above. Underresourcing is due to several factors such as its omission in the project management. There is often a high level of variability as a result of the case-specific nature making it difficult to fit into the established planning structure. Existing project management frameworks, such as the PMBOK [9], are general approaches. While they acknowledge the commissioning stage they do not, and cannot reasonably be expected to, provide case-specific guidance on commissioning. They treat commissioning very lightly and rely on the practitioner to identify whether or not commissioning is an important part of the project. The literature suggests that practitioners too often fail to realise the importance and therefore fail to plan sufficiently. Alternatively, project managers may simply be overly optimistic about the risks associated with the installation of a new system. Whatever the reason, the result can be insufficient resources being allocated, with the consequence of poor completion. Incorporation of a broad conceptual model of commissioning into the project management practices would be the first logical step. Commissioning draws from several project knowledge areas such as integration, communication, and risk management. The logical approach is to incorporate into the project life cycle between the execution and closing phases [4, 6, 8].

Problem Definition

Current models of commissioning tend to be simplistic, and relevant only to specific areas. They are focused on the process and consequently tend to produce a somewhat prescriptive list of tasks that need to be performed. A higher-level reconceptualisation of the commissioning process, with the development of a more general theory, could be valuable.

The purpose of this work is to develop a more holistic and integrative theory of commissioning. The specific emphasis is on reducing process downtime, without compromising plant reliability. This is worth attempting as it has the potential to provide a general framework in which the other more process-specific models can be placed.

APPROACH

We start by reconceptualising commissioning in broad terms. We categorise the commissioning strategies according to the operational risk. This results in three categories: direct, parallel, and advanced. We then apply a system modelling method to embed these within the broader manufacturing context. Finally we apply the new framework to two case studies to demonstrate the applicability.

RESULTS

Categorisation of Commissioning Projects

Starting Premise

We start with the premise that the value of commissioning is essentially one of systematic risk reduction that is, used to minimise the risk associated with the installation of a new piece of equipment. More specifically the application of commissioning for the installation of new equipment into the process industry reduces the risk of equipment damage, environmental health and safety, and process downtime.

Thus commissioning is a strategy for treating risk [24]. This has the further important implication that the treatment, hence type of commissioning, can be aligned with the degree of technical risk that the organisation can accept. Thus we specifically link commissioning, as a treatment strategy, to the concept of "tolerable risk" within the risk management literature, and to the concept of strategic risk for the organisation as a whole [25]. This also has contractual implications in project-setup phase, where there is a need to differentiate between the commissioning risk elements and proportion them between the equipment manufacturer, project management organisation, and plant owner [26].

From this starting assumption we identify three categories of commissioning, as strategies in response to organisational risk-tolerance. These are direct commissioning, advanced commissioning, and parallel commissioning. Each has strengths and weaknesses. They can be deployed individually or together.

Direct Commissioning

Direct commissioning is the classical approach to commissioning where the new equipment is installed and the system must remain offline as commissioning is completed. Direct commissioning is the most straightforward approach as no additional equipment or simulation is required. The new equipment is installed into its operational position and the process cannot restart until the system has been commissioned and is running correctly. There is a high level of downtime in this process as the whole system cannot be operated until the new unit is electrically, mechanically, and operationally tested. There is also the risk of having to reinstall the old unit if there are significant complications at any phase of the commissioning process. Direct commissioning is often reserved for well-established unit operations such as new pumps and flow meters. Direct commissioning is most effective when it is used on well-established system and ones that are not a key requirement of the process.

Advanced Commissioning

Advanced commissioning is the process of operating the new unit in advance of installation and in isolation of the main process operation. Advanced commissioning requires the simulation of all proprietary systems that interact with the new unit. Simulation can be extremely complicated or simple depending on the level of interaction between the process and the new unit. (In this context "simulation" can refer to the artificial provision of physical inputs to the new machine or unit, smaller scale models, and mathematical modelling of the functional behaviour of the unit.) Advanced commissioning allows for the electrical, mechanical, and part of the operational testing to be completed. The full functionality of the unit cannot be proven as the system is being simulated by external means, which will always be an approximation of reality. Advanced process is extremely valuable for the development of new technology as it allows for the verification of novel processes at low risk. The most common type of advanced commissioning is the development of model systems which both simulate the operation of the system and the new unit. Advanced commissioning can also include computer simulation of new process which provides a cost effective method of developing concepts in the early stages of design. Advanced commissioning is valuable at proving conceptual designs of new technology. The main drawback of advanced commissioning is that the process is only simulated so there is still the potential that the unit can fail when installed into its operational environment.

Parallel Commissioning

Parallel commissioning is the testing of the new system in parallel to the operating system. Parallel commissioning is the most rigorous form of physical and operational commissioning. It allows for the new unit to be tested under full operational conditions, with low risk of significant process interruptions due to the added redundancy of the old system present in an operational capacity. However it also has the highest cost as it requires the duplicate hardware systems and additional structural space. The only risk associated with parallel commissioning is the integration between the two systems. Often there is some type of switching or merging component in these systems which may require minor process stoppage for installation. Parallel commissioning is often completed when it is critical that the process must not stop for any extended period of time. It often lends itself to processes with few interactions between new unit operation and the rest of the process. Parallel commissioning is seldom utilised due to the requirement of a process that can accommodate both the new and old unit.

Conceptual Model

Having identified three types of commissioning, we next seek to set those within a conceptual framework. This is worth doing as it has the potential to identify the situational variables relevant to each type of commissioning. This in turn can be used to further build a theoretical foundation, and provide guidance to practitioners.

Approach

The modelling method uses a structured, deductive process to decompose the process being analysed into multiple subactivities (functions) and for each deduce the initiating events, the controls that determine the extent of the outputs, the inputs required, the process mechanisms that are presumed to support the action, and the outputs. The model was then inductively reconciled with elements of the existing body of knowledge on this topic, and successively refined. The end result is a graphical model that describes the relationships between variables, thereby providing a synthesis of what is known and surmised about the topic. The model is expressed as a series of flowcharts using the integration definition zero (IDEF0) notation [27, 28]. With IDEF0 the object types are inputs, controls, outputs, and mechanisms (ICOM) and are distinguished by placement relative to the box, with inputs always entering on the left, controls above, outputs on the right, and mechanisms below.

Develop Production Capability (Prd-1)

The broader context is that commissioning occurs as part of the development of production capability, and our model starts at this level. (This is already the second level into the model; the top level, which is not shown here, includes product design, operation of the plant, control of production flow, quality, distribution to market, packaging, health and safety, lean/JIT, among other activities. However the present paper focuses on the commissioning activities.) See Figure 2. Commissioning is included as element 5 and occurs towards the end of the plant-development process. Other important activities are the following.

- Determine manufacturing/production sequence.
- Design of the production plant, which also includes the plant layout, material handling, plant control and automation, and (for manufacturing) the development of production tooling and flow control, for example, just-in-time (JIT). Analysis of technology risk (9) is another activity associated with the design phase.
- Building the production system (4), and the associated project management activities.
- Decommissioning the plant (7).
- Project management (8). We note the importance of project management methods in supporting many of the activities of commissioning. There are several models of project management

that might be inserted here, including [9, 29], but these are not specific to commissioning and therefore not detailed further at this point.

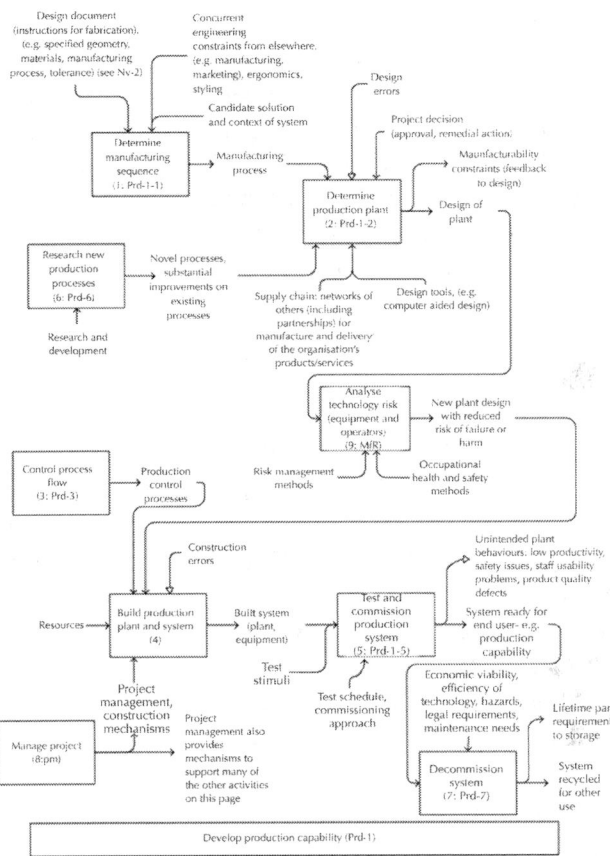

Figure 2: Model for the development of production capability.

We do not deal with these other activities here, but instead move the focus to the test and commission activities. Before doing so, we draw attention to some hollow arrows, which represent errors, in particular design and construction errors, at (2) and (4), respectively, and the possibility for unintended plant behaviours at (5): low productivity, safety issues, staff usability problems, product quality defects, and so forth. This point is important because the commissioning model that follows specifically seeks to address these risks.

Test and Commission Production System (Prd-1-5)

The model for commissioning a new piece of plant equipment is shown in Figure 3 (Prd-1-5). The conventional commissioning process is included here, as are the new concepts for commissioning approach. One of the conventional activities is to verify instrumentation and control systems (1), which involve the systematic checking of installed hardware against plant schematics. The checks are progressively done for connectivity, cold operation, hot operation, and process control. We do not detail those processes here and instead refer the reader to source material [5] which has information that is useful to practitioners. The final objective of commissioning is also well known, to deliver an operational production system (6) that is ready for the client to use.

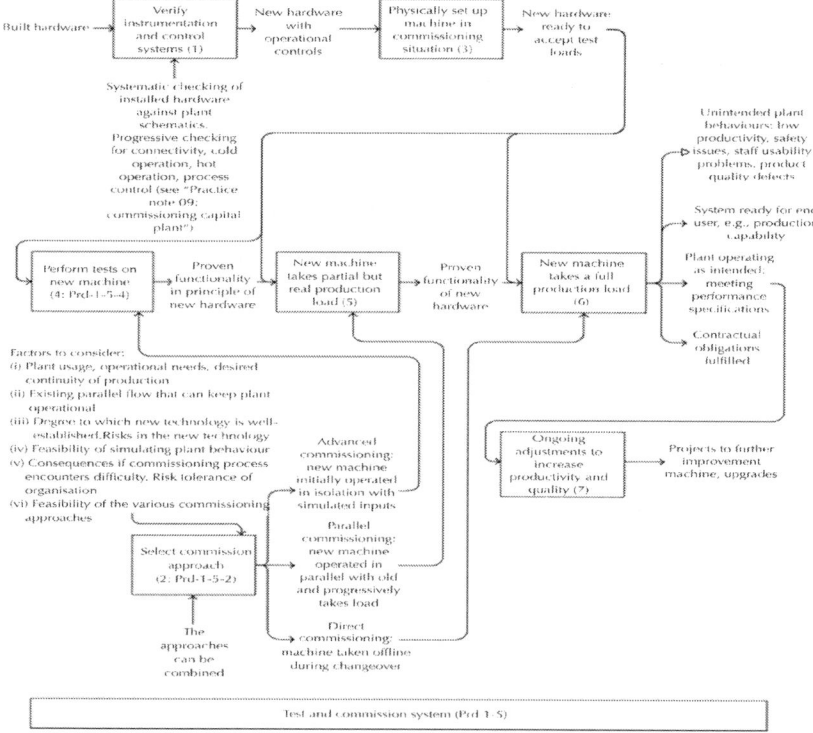

Figure 3: Model for the test and commissioning activities.

Where our model differs is the inclusion of a deliberate stage of deciding which of three commissioning approaches to use in the situation (2): direct, advanced, or parallel. We also note in passing that the quality and lean imperative for continuous improvement will generally mean that there will be ongoing adjustments to increase productivity and quality (7) after the machine has been commissioned. Thus commissioning the machine and closing the contract with the client may be the end of the involvement of the machine builders, but are not the end of the life cycle for the machine itself. This again has contractual implications in the form of service and warranty support from the vendors, and reliability centred maintenance by the plant operators. There is also the decommissioning to consider, which can be a project in itself. (In extreme cases, e.g., nuclear power plant, the cost of decommissioning is comparable to the initial construction cost. If there has been a catastrophic failure of the plant then the decommissioning cost can vastly exceed the construction cost.)

Select Commission Approach (Prd-1-5-2)

The decision involves a choice of direct, advanced, or parallel commissioning. These are not mutually exclusive. Instead some of them may be done sequentially, as shown in Figure 4. For example, advanced commission may precede either of the other two.

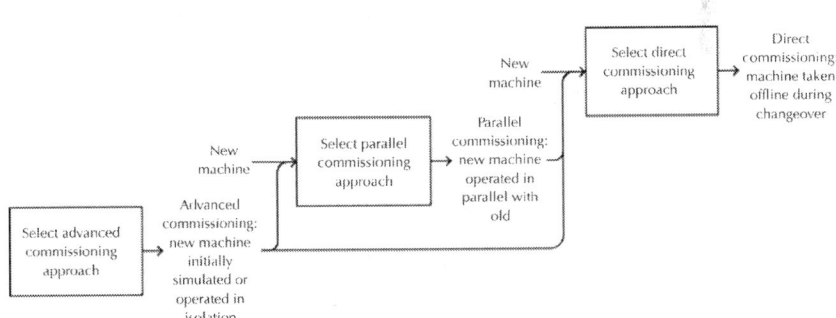

Figure 4: Relationship between the three commissioning approaches: advanced, parallel, and direct.

The various factors relevant to this commissioning decision are anticipated and clustered in groups: ability to recover from a failed installation, assessed or perceived technology risk, desired operational continuity, and timing considerations. The detailed model and the factors within each cluster are shown in Figure 5. At this stage the model is primarily logical and qualitative and is intended as a debiasing tool and a guide to action rather than a decision algorithm. It is also a framework for further research in that it proposes subjective relationships of causality that can subsequently be tested and developed as appropriate. (It may even be that in certain areas it could be possible to develop a mathematical model to support the decision, particularly in well-defined areas. Specifically, the model incorporates risk assessment and it is not impossible that there could be well-defined situations where the variables can be determined with sufficient precision that a quantitative risk assessment coupled with (say) a Boolean consideration of the other factors might make for a sufficient mathematical model. However further research would be required to take it to this level of detail.)

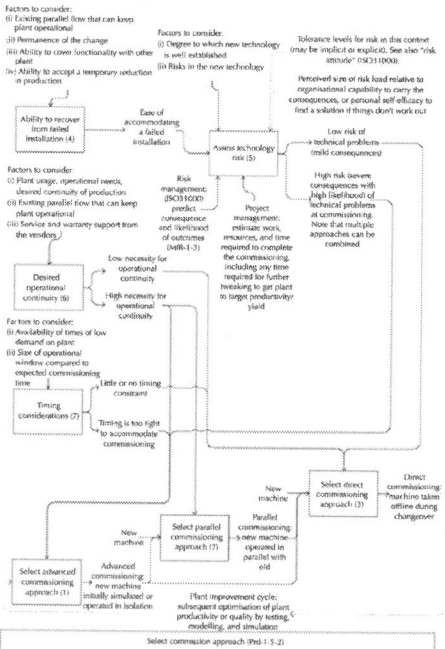

Figure 5: Factors relevant to the commissioning decision.

Thus the model proposes that the following decision factors are relevant.

- Advanced commissioning is appropriate where technology risk is high, operational continuity is required, and timing constraints are tight.
- Parallel commissioning is appropriate where operational continuity is required and timing constraints are tight.
- Direct commissioning is appropriate where technology risk is low, operational continuity can be disrupted, and timing constraints are loose.

Finally, to complete this part of the conceptual framework, a model is provided for the testing activities of commissioning; see Figure 6.

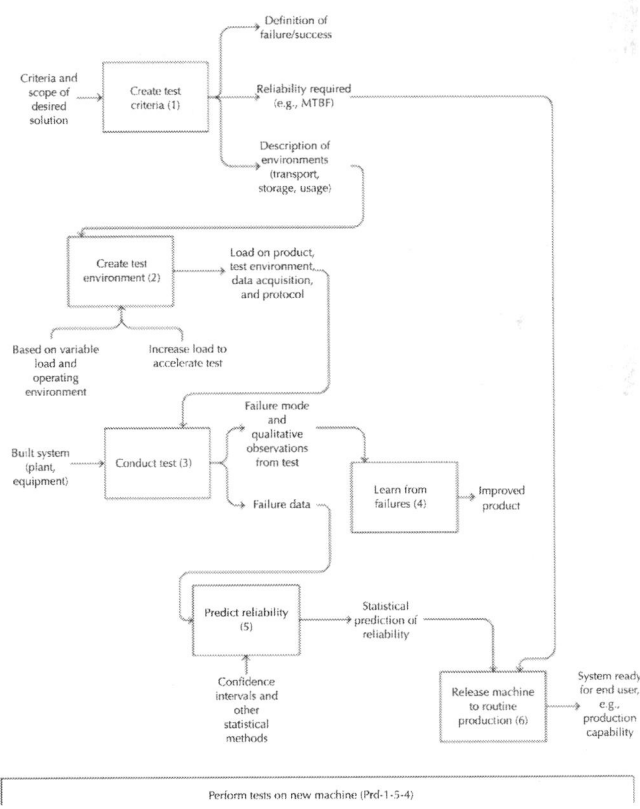

Figure 6: Model for the test activities.

CASE STUDIES

Two cases studies were completed to determine the relevance of this commissioning model in the process industry. First was the development of a novel vertical screw system in the fertilizer industry which used the advanced approach to commissioning. Second was the installation of a ship unloader in the aluminium industry which used a parallel approach to commissioning.

Vertical Screw Project

Ballance Agri-Nutrients single superphosphate plant at Awarua (New Zealand) had recently designed and installed a new phosphate rock feed system. A new vertical screw system was developed to replace the old gravity feed vertical chute which was prone to blockages in the highly reactive and humid environment present in the reaction chamber. The vertical screw was designed to increase the reliability of the process by forcing the rock into the reaction chamber, hence reducing the number of blockages. An advanced approach to commissioning was completed for this new project due to the risk associated with the installation of a complex and untested unit into a critical position in the production line.

The advanced approach to commissioning allowed for rigorous testing to be completed in a controlled environment with low risk to production capacity. Commissioning was completed in two stages with the development of a model system and full scale commissioning of the new system before installation.

Development of the model system allowed the basic concept of the system to be tested. The model was constructed out of crude materials and was tested under a range of conditions to determine the optimal operating parameters. The full scale vertical screw was designed and constructed based on the results obtained from the model system.

Commissioning was completed by operating the system under a set of conditions established to simulate normal operation. The simulation was completed by assembling the new system directly adjacent to its intended future position in the operational plant. It was wired into the system using all of the final wiring components but was not installed into the process. A feed hopper was fitted to the inlet of the screw to simulate the priority feed system and water and various other components were used to simulate the environment of the reaction chamber. The operation of the vertical screw during the advanced commissioning phase can be seen in Figure 7. The process of running the system under simulated operating conditions allowed the full commissioning of the mechanical and electrical systems. It also allowed for the partial commissioning of the operation capacity. During this process it was found that several components did not operate as expected. Changes were made to the system and were re-commissioned without any negative effect to the production capacity of the plant.

Figure 7: Advanced commissioning of vertical feed screw during plant operation. Image shows plant being fed with material via a temporary arrangement while being commissioned. (Photograph by K Lawry and used by permission of Ballance Agri-Nutrients.).

The process of commissioning prior to the installation of the vertical screw into the system was extremely successful. Full mechanical and electrical commissioning was completed as the full electrical system was used to drive the advanced commissioning process. The operational testing identified several design flaws in the vertical screw. It also reduced the uncertainty associated with the operation of this new technology.

Advanced commissioning has one significant drawback; it only reduced risk associated with the installation of the new vertical screw. The simulated process cannot represent the real process exactly. There are several factors such as continuous operation in the highly reactive environment that cannot be tested until the system is fully installed. Nonetheless the advanced approach was effective in eliminating latent defects and thereby reduced the overall risk.

Tiwai Point Ship Unloader

New Zealand Aluminium Smelters (NZAS) in Bluff installed a new ship unloader on the Tiwai wharf for the unloading of alumina and coke from incoming ships. The new unloader produced by Alesa Engineering Ltd. replaced the forty-year-old unloader that was installed when the smelter was first constructed in 1971. The new unloader has a significantly increased capacity capable of discharging at 1,000 tonnes per hour (TPH) of alumina and 600 TPH of coke compared to the old unloader that was only capable of discharging 235 TPH of alumina and 250 TPH of coke [30]. The new ship unloader was installed with the aim of reducing the time required for a ship to spend unloading. Less time spent unloading will mean a more efficient use of shipping resources and the reduction of costs associated with slow turn around (demurrage).

Installation and commissioning was completed by Alesa under the guidance of specialist project engineering company Bechtel who work onsite at the Tiwai point aluminium smelter. This process had to be completed under tight constraints as it was critical that there were no process interruptions. The smelter is a continuous process which cannot be shut down or restarted without high associated cost.

It was decided to take a parallel approach to commissioning; the old unloader must remain in a fully operational state until the new system has been thoroughly commissioned and proven capable of carrying the full operational load.

Keeping the old ship unloader in an operational state significantly reduced the risk of supply interruptions, but introduced additional concerns relating to the integration of the two systems. Integration of both unloaders on the same wharf was completed by limiting the operation of the old unloader to the north half of the wharf, while the new unloader was installed and commissioned on the south half, as seen in Figure 8. Limiting the old unloader to half of the wharf increased the unloading time as the ship had to be manoeuvred around to allow access to all of the holds. This was taken as a minimal sacrifice to ensure consistent supply.

Figure 8: Parallel operation of old (left) and new (right) ship unloaders on the Tiwai wharf. (Photograph by K Lawry.).

The main modification that was required for the integration of the new unloader into the existing infrastructure was the installation of a new conveyor system to replace the southern half of the existing wharf conveyor, clearly this has cost implications. This was completed before the new unloader was installed. Both the new and old conveyor systems operated as one continuous conveyor that serviced the new and old unloaders simultaneously. The upgrade of the conveyor acted to integrate the two unloaders into the overall process. The new conveyor and ship unloader were constructed and assembled off site. The units were then transported and lifted into place. The use of pre-constructed assembled units significantly reduced installation time, therefore allowing installation to be completed in the short window between scheduled shipping movements.

Several complications emerged during the commissioning, and these extended the project duration. However the parallel approach meant that these had no impact on the overall production capacity of the smelter. The reduction in unloading capacity caused by the limitation of the old unloader to the north half of the wharf was quickly offset by the high capacity of the new unloader even when it was operating at a reduced output. Parallel commission proved to be a successful method of commissioning as there were relatively minor additional costs and the risk of process downtime was completely mitigated. The old unloader was decommissioned once the new unit was fully operational [31].

DISCUSSION

This paper makes several novel contributions. First, it provides a novel conceptual framework for the commissioning process. The model represents the decision making within the process, the broader context in which plant commissioning occurs, as well as making provision for the finer details. The novelty is creating a structured representation of the commission process, where models are otherwise sparse. Commissioning is generally an ad hoc process, and the value of this new framework is that it provides a structured theoretical foundation for this important activity.

A second contribution is the categorisation of commissioning into three main types: advanced, parallel, and direct. This exposes the variability of strategies within the commissioning process, so it becomes apparent that there is not merely one universal approach to commissioning. Achieving this adds choice to the project planning. It makes it clear that there are choices that practitioners can make, and stating these choices encourages a thoughtful consideration of the planning and resource implications thereof. This categorisation thus adds richness to the conceptual model and makes the decision points more explicit, without being prescriptive.

A third contribution is the development of a model for use by practitioners. The model captures and represents the proposed situational variables (contingency factors) involved in the process. This is valuable for informing the decision making of practitioners. The applicability of the model has been demonstrated by case studies.

A fourth contribution is the integration of commissioning into other management models. The model provides integration with the "risk management" and "project management" disciplines. This is valuable because it shows practitioners how commissioning may be approached in a more holistic manner. The commissioning model is also integrated into a wider model for the development of production plant, and thereby into "manufacturing engineering" including quality and lean manufacturing. Space does not permit full description of this integration, but the point is that the work shows that this integration is indeed feasible. The model is represented in IDEF0 notation, which is a production engineering notation, meaning that it is readily comprehendible in that context.

Overall what has been achieved is to replace the otherwise ad hoc process of commissioning with a systematic process complete with proposed decision factors and internal models of causality. There are implications for practitioners in the model, in the form of flowcharts identifying the critical success/failure factors for commissioning. Thus tentative recommendations can be made for the best commissioning approach for a given situation.

There are also implications for further research. The model is at least partly conjectural, and further research could be directed at establishing the validity of the proposed causal relationships. Another strand of research could be directed at further refinement of the model, and its extension deeper into specific cases, that is, further investigation of the situational variables.

CONCLUSIONS

Commissioning is extremely valuable to all projects but is poorly defined in the project management body of knowledge. The existing literature on commissioning is focussed on specific tasks, and holistic perspectives are lacking. This work has reconceptualised commissioning and shown that it is possible to identify three main types of commissioning (direct, parallel, and advanced) and construct a generalised conceptual framework around them. This approach to commissioning has been demonstrated by application to case studies.

The value of this work is that it presents a different and more structured way of thinking about commissioning. This allows for a more critical examination of how to complete the commission for a particular project, and ultimately the potential for a better commissioning outcome for practitioners. For theorists the benefit is that a generalised model has been developed, thus a foundation for future advancement of the subject. We have shown that the commissioning activities can be integrated into the risk management, project management, and production engineering bodies of knowledge.

ACKNOWLEDGMENTS

The authors would like to thank Ballance Agri-Nutrients and Richard Sweney at New Zealand Aluminium Smelters for providing the information required for the cases studies. These cases studies provided a valuable insight into how commissioning is completed in industry and would not have been possible without the help from these organisations.

REFERENCES

1. R. Bernhardt, "Approaches for commissioning time reduction," Industrial Robot, vol. 24, no. 1, pp. 62–71, 1997.

2. R. B. Brown, M. B. Rowe, H. Nguyen, and J. R. Spittler, "Time-constrained project delivery issues,"AACE International Transactions(PM. 09), pp. 1–7, 2001.

3. P. Gikas, "Commissioning of the gigantic anaerobic sludge digesters at the wastewater treatment plant of Athens," Environmental Technology, vol. 29, no. 2, pp. 131–139, 2008.

4. D. Horsley, Ed., Process Plant Commissioning, Institution of Chemical Engineers, Rugby, UK, 2nd edition, 1998.

5. IPENZ, Practice Note 09: Commissioning Capital Plant, IPENZ, Wellington, New Zealand, 2007.

6. J. Kirsilä, M. Hellström, and K. Wikström, "Integration as a project management concept: a study of the commissioning process in industrial deliveries," International Journal of Project Management, vol. 25, no. 7, pp. 714–721, 2007.

7.   NHS, Project Management in a PCT Environment, National Primary and Care Trust Development Programme, 2004.

8.   B. Peachey, R. Evitts, and G. Hill, "Project management for chemical engineers," Education for Chemical Engineers, vol. 2, no. 1, pp. 14–19, 2007.

9.   Project Management Institute (PMI), A Guide to the Project Management Body of Knowledge (PMBOK Guide), Project Management Institute, Newtown Square, Pa, USA, 4th edition, 2008.

10. P. V. Thomas, "Best practice for process plant modifications (fertilizer plants)," Cost Engineering, vol. 45, no. 5, pp. 19–29, 2003.

11. V. S. Sohmen, "Capital project commissioning. Factors for success," in Proceedings of the 36th Annual Transactions of the American Association of Cost Engineers (AACE' 92), Orlando, Fla, USA, June-July 1992.

12. H. M. Guven and S. T. Spaeth, "Commissioning process and roles of pyers," in Proceedings of the ASHRAE Winter Meeting, la, New Orleans, La, USA, January 1994.

13. S. K. Shome, "Integration of commissioning activities in project management in power sector," inProceedings of the Project Management in the Power Sector Seminar, Ooty, India, November 1982.

14. M. G. Tribe and R. R. Johnson, "Effective capital project commissioning," in Proceedings of the 54th IEEE Pulp and Paper Industry Technical Conference (PPIC' 08), Piscataway, NJ, USA, June 2008.

15. E. E. Choat, "Implementing the commissioning process," in Proceedings of the Winter Meeting of ASHRAE Transactions, Part 1, Chicago, Ill, USA, January 1993 1993.

16. S. Doty, "Simplifying the commissioning process," Energy Engineering, vol. 104, no. 2, pp. 25–45, 2007.

17. E. Schepers, "Commissioning chemical process plant," in Proceedings of the 2nd National Chemical Engineering Conference, pp. 60–69, Institution of Chemical Engineers, University of Queensland, Surfers Paradise, Australia, 1974.

18. G. Shimmings, "Reflections on the causes of delays in commissioning automated materials handling projects," in Proceedings of the 3rd International Conference on Automated Materials Handling, Birmingham, UK, 1986.

19. A. Levi and M. Stonell, "Project management and commissioning of industrial boiler plant," Institution of Mechanical Engineers, Conference Publications, pp. 55–67, 1979.

20. E. Cagno, F. Caron, and M. Mancini, "Risk analysis in plant commissioning: the Multilevel Hazop,"Reliability Engineering and System Safety, vol. 77, no. 3, pp. 309–323, 2002.

21. V. Ramnath, "How you can precommission process plants systematically," Hydrocarbon Processing, vol. 90, no. 4, pp. 119–124, 2011.

22. A. Rautenbach, "Site acceptance testing and commissioning of process control systems," Elektron, vol. 19, no. 5, pp. 40–44, 2002.

23. G. Reid, "How to achieve successful startup and commissioning for instrumentation and controls project," in Proceedings of the Advances in Instrumentation and Control Conference, vol. 47, pp. 121–124, ISA Services, Houston, Tex, USA, October 1992 1992.

24. ISO 31000, Risk Management—Principles and Guidelines, International Organization for Standardization, 2009.

25. D. J. Pons, "Strategic risk management in manufacturing," The Open Industrial and Manufacturing Engineering Journal, vol. 3, pp. 13–29, 2010.

26. J. Leitch, "Eliminating the risks to starting up your plant right the first time," Hydrocarbon Processing, vol. 85, no. 12, pp. 47–52, 2006.

27. FIPS, "Integration definition for function modeling (IDEF0)," 1993,http://www.itl.nist.gov/fipspubs/idef02.doc.

28. KBSI, "IDEF0 overview," 2000, http://www.idef.com/idef0.htm.

29. D. J. Pons, "Ventures of co-ordinated effort," International Journal of Project Organisation and Management, vol. 4, no. 3, pp. 231–255, 2012.

30.   NZAS, "Unloader," in Tiwai Pointer, pp. 1–7, Newsletter of New Zealand Aluminium Smelters, 2011.

31.   NZAS, "A look at our new ship unloader," in Tiwai Pointer, pp. 1–7, Newsletter of New Zealand Aluminium Smelters, 2012.

Chapter 3

Chemical Analysis on Mongolia's Natural Bitumen

Erdenetsogt Bat-Erdene[1], Batdelger Byambagar[1],
Erdenee Enkhtsetseg[1], and Budeebazar Avid[2]

[1]School of Material Science of the Mongolian University of Science and Technology, Ulaanbaatar, Mongolia

[2]Institute of Chemistry and Chemical Technology, Mongolian Academy of Science, Ulaanbaatar, Mongolia

ABSTRACT

To extract pure bitumen, the bitumen from Bayan-Erkhet, Zuunbayan and Ukhaa was prepared into small particles of 0.2 - 0.5 cm and then it was infused with chloroform in the Soxhlet apparatus [1] [2]. The physical-mechanical properties were identified after the solvent was extracted from the chloroform infused bitumen through the vacuum evaporation method. The characteristics of the debris without bitumen or the remains after the infusion were examined in details. The hydrocarbon content of the bitumen was identified with the device: Agilent 7890-5975c Gas chromatography mass spectrometer.

INTRODUCTION

Our nation imports bitumen that is commonly used for constructions and asphalt concrete road coverings. Within the recent years, the prices of raw oil and its related products have been increasing significantly, and the supply of it has become more limited. Hence, many countries in the world started seeking new materials and sources to substitute and use for road, constructions, fuel, energy and industries, and the researches for these are expanding significantly. It is recorded that our country has substantial amount of bitumen and oil shale, and tentatively there are about 800 million tons of deposits in 60 minefields [3]. Thus, we need to use new technological methods to produce and explore high quality products such as high quality roads, bitumen for raw construction materials and fuels and other related products.

There have been various studies on the bitumen and oil shale minefields, the geological formations and the deposits. Unfortunately, there have not been sufficient studies on the chemical compositions, structures, and chemical-technologies done.

MATERIALS AND METHODS

Materials

The materials used for this work include:

- Bayan-Erkhet deposits of Tuvaimagandsum center, located in the south east, 40 km and 200 km southeast of Ulaanbaatar Railway 14th cross roads 55 kilometers to the northeast. The deposit has reserves of 1.2 million tons of bitumen sands [4].

- Zuunbayan Dornogovi bitumen sand deposit is located 50 km south across the petroleum deposit consists of a 4.2 km by 2 small deposits and reserves of 330 tons [3] [5].

- Zuunbayan   Dornogovi   Ukhaaofbitumensand   deposit sinthecurrentgeologicalexplorationwork

Methods

Sample Preparation

Preparing the samples between an area different positions from the supply after to crush 1.25 mm diameter prepared fedsieves.

Bitumen Separation Method

40 times the volume of bitumen dissolved in hexane divided asphaltene. The differentiated asphalt from the Maltese parts were condensed in the Soxhlet apparatus with ASK type activated Silica gel and the butyric compounds with i-hexane, tar-like compounds with 1:1 ratio alcohol:benzol alternatively [6] [7] . Test are shown in Table 1.

Mineral Group Methods

Mineral part of the particle module, shares spacing, the actual density parameters such as MNS 392:98, MNS 392:98, and MNS 2916:2002 standard specifies methods [6].

Bitumen Hydrocarbon Composition Method Determination

The hydrocarbon content of the bitumen was identified with the device: Agilent 7890-5975c Gas chromatography mass spectrometer Test conditions: GC: Carrier gas: 99.999% He; Inlet: 300°C; Transmission line: 280°C; Column: HP-5MS fused silica capillary column (60 m × 0.25 mm × 0.25 μm); Column temperature: Initial temperature 50°C, 1 min; 15°C/min heating to 120°C, and then to 3°C/min up to 300°C to maintain 25 min; Carrier gas flow: 1 mL/min. MS: EI, 70 eV; Full scan [8] .

Pure Bitumen Test Methods

Physical and mechanical characteristics of bitument MNS 5109-2001, MNS 5211-2002, MNS 5110-2001, MNS 328-2000, MNS AASHTO T40-2003 standard specifies methods [6]

RESULTS AND DISCUSSION

The segregated bitumen from the Bayan-Erkhet, Zuunbayan and Ukhaa minefield bitumen's general parameters are shown in Table 1.

From the above table, the pure bitumen's yields of the bitumen of Bayan-Erkhet, Zuunbayan and Ukhaa infused with chloroform 14.75%, 15.84%, 10.86%. When the pure bitumen content in natural bitumen is between 10% - 15%, it is considered to be an economically beneficial raw material [11]. Compared to oil bitumen, the natural bitumen has more surface active compositions (tar, asphaltogen acid, and its' anhydride), and they go through adsorption at the mineral parts and the ability to bond with the mineral parts enhances [12]. Some technical properties, like ash-like, humidity, volatile substance contents of the bitumen sand sample and the extracted mineral sample, are identified in Table 1. From the table, it can be observed that the humidity content (0.72, 0.64, 0.89 mas. %) is much lower than coal, shale and other hard organic raw materials. This might be due to the bitumen's "hydrophobic" quality, to push water and not dissolve in it. As bitumen is considered to have low organic substance content, it has high ash content (73.41, 79.32, 83.24 mas. %); however, volatile substance content (22.52, 19.84, 16.32 mas. %) indicates that the parts that make up the organic mass has volatile characteristics. Mineral parts are the remains of the chloroform infused bitumen, and the yield of the volatile substance is 0.79, 0.43, 0.64 mas. %, which indicates that the organic parts are almost completely 100% infused with the chloroform. Another proof is the ash-like content of the minerals (99.13, 98.35, 99.32   s. %).

Table 1: Bitumen's general parameters

№	Parameters	Bayan-Erkhet [9]		Zuunbayan		Ukhaa [10]	
		Sample	Within the minerals	Sample	Within the minerals	Sample	Within the minerals
1	Bitumen yield, mas. %	14.75		15.84		10.86	
2	Humidity, mas. %	0.72		0.64		0.89	
3	Volatile substance, mas. %	22.52	0.79	19.84	0.43	16.32	0.64
4	Ash-like, mas. %	73.41	99.13	79.32	98.51	83.24	99.32

The following compounds were identified in the natural bitumen that we are studying: (Table 2).

The naphtene's total hydrocarbon content in Bayan-Erkhet, Zuunbayan, Ukhaa's natural bitumen is higher than in saturated hydrocarbons. As the bitumen from the above mentioned minefields have high asphalt contents, it is considered to have "Gel" bitumen [12]. Most natural bitumen's tar content is higher and the oil content is lower than oil bitumen, which allows them to be more stable. When tar content is higher, the tar in asphalt's dissolving characteristics "lyophilization" increases, and a stable colloid solution is formed. On the other hand, if the oil content is high, it has negative impact on bitumen quality, as the asphalt's oil does not dissolve "lyophobic" [13]. From the printed materials, it can see that oil bitumen's main components, high molecular tar, asphalt compounds, ratios are 1:1 and the total content should be close to 50%. However, the remaining 50% consists of oily (fatty) compounds [12].

Important characteristics of the pure bitumen that was separated from the bitumen sand through the standard methods are shown in Table 3.

The pure bitumen extracted from the Bayan-Erkhet bitumen sand is in the category of thick bitumen because of the needle sinking depth (48.5 mm at 25°C), andhydrocarbon content (tar content is relatively high 60 mas.%). From the technical results, it can be seen that majority of the physical-mechanical parameters, including needle sinking depth at 25^0, temperature to soften, density is within the technical requirements of 40/60 road bitumen. However, the elasticity is lower than the standard, which can be explained by the low asphalt content. On the other hand, the pure bitumen extracted from Zuunbayan bitumen sand is in the category of liquid bitumen due to the needle sinking depth (223 mm at 25°C) and oil content (42.5 mas.%). The physical-mechanical properties, such the needle sinking depth at 25°C, temperature to soften anddensity, are within the 200/300 type road bitumen's technical requirements [1]. The pure bitumen extracted from Ukhaa bitumen sand is relatively close to Zuunbayan's bitumen sands properties and it is also within the 200/300 type bitumen.

Table 4 below shows the physical-mechanical properties of the segregated mineral parts of the bitumen sand.

Sand particle modules are categorized into:
- Large particle > 3.1 mm.
- Medium particle 2.1 - 3.1 mm.
- Small particle < 2.1 mm.

From the results, Ukhaa, Bayan-Erkhet's bitumen sand's sand belongs to medium sized particle category, which indicates that these can be used for asphalt concretes, but the dust clay particles in the sand is higher than the accepted technical requirement. On the other hand, Zuunbayan bitumen sand's sand belongs to small sized particle module.

Depending on the strainer diameter size, the samples are called:
- Dust, clay for parts that went through 0.075 mm diameter strainers.
- Sand for parts that passed through 0.075 - 4.75 mm diameter strainers.
- Rock for parts that pass through or bigger strainers with diameters of 4.75 mm [6].

Thus, from the study results, Bayan-Erkhet, Zuunbayan's sands were 8.73, 10.66%, which indicates that the smaller parts were comparatively more.

Table 2: Group compositionofbitumen

Parameters	Deposits		
	Bayan-Erkhet	Zuunbayan	Ukhaa
Saturated HC: ¾ n-Alkane ¾ Cycloalkane Total	10.81 16.89 27.70	11.42 20.38 31.80	7.38 20.49 27.87
Aromatic HC: ¾ Alkyl aromatic ¾ Polynuclear aromatic Total	6.22 8.17 14.39	6.60 11.82 18.42	5.85 9.48 15.33
¾ Asphaltene	28 .42	25 .66	24 .04
¾ Tar	18 .71	20 .18	18 .12

HC-hydrocarbon.

Table 3: Characteristics of bitumen

№	Parameters	Bayan-Erkhet	Technical Requirement 40/60	Zuunbayan	Ukhaa [10]	Technical Requirement 200/300
1	Penetration at 25°C	48.5	41 - 60	223.3	274	201-300
2	Not softening and ring	47.7	40 - 50	34.5	26.2	>35
3	Ductility at 25°C	36	>100	57.5	-	-
4	Temperature to flame, °C	193	> 220	190	210	>200
5	Density, g/cm³	1.03	0,95-1,15	0.98	0.979	-

Table 4: Bitumen sand minefield's mineral part studies

№	Parameters	Technical requirement	Ukhaa	Bayan-Erkhet	Zuunbayan
1	Sand particle module	2.1 < 3.1	2.21	3.12	1.6
2	Dust clay content, mas. %	<3	19.97	15.46	20.13
3	Density, g/cm³	>2.4	1.559	2.58	2.55
4	0.075-antecedent, %	-	1.78	8.73	10.66
5	Organic mix. %	no	no	no	no
6	Flow, flexibility index	unarticulated	unarticulated	unarticulated	unarticulated

The activity levels of natural radioactive isotopes, element contents, and Rad's equivalents of the cenosite ashes of Bayan-Erkhet, Zuunbayan and Ukhaa bitumen sand minefields were identified through gamma spectrometer method (Table 5).

The maximum allowed radioactive Radon for construction material production is defined to be 370 Bq/kg. The above mentioned minefield's Radon equivalent dose in cenosite parts of the ashes are almost three times lower than the standard; hence, it can be directly used for road and other construction materials.

CONCLUSIONS

From the results of this work, the following conclusions can be drawn:

- The bitumen categories of Bayan-Erkhet, Zuunbayan and Uhaa bitumen sand minefields were identified. For instance: Bayan-Ekhet's bitumen type was thick viscose "asphalt", whereas Zuunbayan and Ukhaa's bitumen was liquid type. According to the results, Bayan-Ekhet, Zuunbayan and Ukhaa's bitumen sand minefields's natural bitumen has high organic mass yields and tar-like asphalt contents.

- From bitumen mineralogy results of the study minefields, the segregated sands of Ukhaa and Bayan-Erkhet bitumen are categorized into medium particle sized sand, which is suitable to be used for asphalt concretes; however, the dust, alluvium content is higher than the standard. The sand extracted from the

Zuunbayan bitumen sand is in the small particle sized category. From the cenosite radioactivity tests of the above mentioned mine fields, it has been identified that it can be used directly towards road and construction materials.

Table 5: Content of radioactive elements

№	Sample	Isotopic activity, Bq/kg			Contains elements			Rad's equivalent dose, Bq/kg
		^{226}Ra	^{232}Th	^{232}K	U, g/t	Th, g/t	, g/t	
1	Bayan-Erkhet	10	10	1080	0.8	2.4	3.2	121.1
2	Zuunbayan	13	11	974	1.0	2.7	3.3	110.4
3	Ukhaa	18	16	956	1.5	4.0	3.6	115.1

Bayan-Erkhet bitumen has similar physical-mechanical properties as 40/60 type bitumen. This bitumen can be directly used for road coverings. Zuunbayan and Ukhaa's bitumen has similar properties as 200/300 type bitumen for needle sinking depth, temperature to soften and density, and belongs to liquid type bitumen. Hence, these need to be further processed and the structures need to be changed, in order to get high quality bitumen that can be used for roads for the climatic conditions of our country.

REFERENCES

1.   Mongolian National Standard (2007) Road Sector Standards Compilation. Part 3. Organic Adhesive Asphalt. Official Media. Standards and Metrology Center-UB, 54-72.

2.   Abryutina, N.N., Abushaeva, V.V. and Arefev, O.A. (1984) Modern Methods of Investigation of Oil. L., Nedra, 105-106.

3.   Munkhtogoo, L. and Baatar, L. (1991) Report for Exploring Raw Materials for Road and Construction Materials in Dundgobi, Dornogobi, and Tuw provinces. UB, Mongolia.

4.   Munkhtogoo, L., Stone, B. and Dashdondov, J. (1986) Bayan-Erkhet Deposits of Asphalt Survey Report Conducted. UB.

5.   Khongor, O. and Usukhbayar, L. (1985) The Results of Prospecting and Exploration Works Carried Out at the Field on the Sandstones of Zuunbayan , East Gobi Aimag. UB, Mongolia.

6.   Arivjikh, T., Ochirbat, S., Dashzeveg, D. and Lkhagvajav, Ch. (2005) Road-Building Materials Science. UB, 331-397.

7.   Abryutina, N.N., Abushaeva, V.V. and Arefev, O.A. (1984) Modern Methods of Investigation of oil. L., Nedra, 105-106.

8.   Terentev, P.B. (1979) Mass spectrometry of Organic Chemistry. M, 223.

9.   Bat-Erdene, E., Tuya, M. and Khulan, B. (2009) Getting a Synthetic Crude Oil from Tar Sand Deposits of Bayan-Erkhet (Mongolia). Chemistry of Oil and Gas Conference Materials. Tomsk, 2009. (The 7th International Conference), 749-753.

10. Bat-Erdene, E., Byambagar, B., Narantsetseg, M., Enkhtsetseg, E. and Avid, B. (2013) Results from Ukhaa Bitumen Sand Minefield. ICCST-2013, The 4th International Conference on Creative Science and Technology, Darkhan, 4-5 October 2013, 415-417.

11. Patel, S. (2007) Canadian Tar Sands-Favorable Opportunities, Technologies and Problems. Hydrocarbon Processing, 87-93

12. Gun, P.B. (1973) Petroleum Bitumen M. Chemistry, 6-71.

13. Yoshida, R., Yoshida, T., et al. (1983) Comparison of the Chemical Structure of Coal Hydrogenation Products, Athabasca Tar Sand Bitumen and Green River Oil Shale. Fuel Processing Technology, 7, 161-171. http://dx.doi.org/10.1016/0378-3820(83)90034-6

Hydrodynamic Cavitation-Assisted Synthesis of Nanocalcite

Shirish H. Sonawane[1], Sarang P. Gumfekar[1], Kunal H. Kate[1], Satish P. Meshram[1], Kshitij J. Kunte[1], Laxminarayan Ramjee[1], Candrashekhar M. Mahajan[1], Madan G. Parande[1], and Muthupandian Ashokkumar[2]

[1]Nanoscience and Engineering Research Group, Department of Chemical Engineering, Vishwakarma Institute of Technology, 666-Upper Indira Nagar, Pune-411037, India

[2]School of Chemistry, University of Melbourne, Melbourne VIC 3010, Australia

ABSTRACT

A systematic study was made on the synthesis of nanocalcite using a hydrodynamic cavitation reactor. The effects of various parameters such as diameter and geometry of orifice, CO_2 flow rate, and $Ca(OH)_2$

concentration were investigated. It was observed that the orifice diameter and its geometry had significant effect on the carbonation process. The reaction rate was significantly faster than that observed in a conventional carbonation process. The particle size was significantly affected by the reactor geometry. The results showed that an orifice with 5 holes of 1 mm size resulted in the particle size reduction to 37 nm. The experimental investigation reveals that hydrodynamic cavitation may be more energy efficient.

INTRODUCTION

The effect of acoustic cavitation on different chemical reactions is well established. Gedanken [1] has reviewed the use of sonochemistry for fabrication of inorganic nanomaterials of various shapes, size, structure, and phases. Acoustic cavitation in liquids leads to two of major effects: physical (streaming, turbulence, microjet, shear, etc.) and chemical (radical production). While acoustic cavitation-induced chemical reactions have been successfully achieved, hydrodynamic cavitation is found to be efficient for applications involving continuous processing such as industrial carbonation operation. It is expected that hydrodynamic cavitation would increase the rate of carbonization reaction by lowering the mass transfer resistance. Hydrodynamic cavitation, in which a liquid is passed through constrictions, such as orifice plate or Venturi, has been found useful in specific chemical reactions. Hydrodynamic cavitation occurs due to the changes in the pressure of liquid flow in a pipe fitted with orifice or Venturi. A liquid experiences a sudden drop in pressure at downstream resulting in the collapse of formed cavities. The collapse of the cavities generates highly reactive radicals, which are responsible for specific chemical reactions. In gas-solid reactions, the dissolution of solids is enhanced due to the turbulent mixing generated by hydrodynamic cavitation. The vigorous mixing enhances the transport of gas solutes to the solid surface that results in an increase in the mass transfer and hence the overall reaction rate [2–4].

Hydrodynamic cavitation has been found useful in the hydrolysis of fatty oils [5] and polymer solutions [6] and in the formation of styrene butadiene rubber nanosuspensions [7]. Morison and Hutchinson [8] have shown the limitations of the Weissler reaction as a model reaction for measuring efficiency of hydrodynamic cavitation. Senthil kumar et al. [9] and Moholkar et al. [10] have reported that the generation of cavities in a hydrodynamic reactor is very much dependent upon the design and the geometry of the reactors. Gogate and Pandit [11] have reviewed the effect of hydrodynamic cavitation on different industrially important reactions, such as the oxidation of toluene, xylene, and transesterification. Suslick et al. [12] studied dependence of tri iodide formation rate on the hydrodynamic pressure used to induce cavitation. Find and Moser [13] have reported on the hydrodynamic cavitation-assisted synthesis of Cu-Zn-Al metal oxides using a mixture of aqueous Na_2CO_3 and NaOH solutions by precipitation.

Calcium carbonate synthesis is one typical example of the carbonation process, in which CO_2 gas is bubbled through $Ca(OH)_2$ slurry, which is also known as "reactive crystallization process." Lin et al. [14] studied the mass transfer effect for the reactive crystallization of calcite crystals in the presence of sodium tripolyphosphate in a multiphase system. The effect of ultrasound on the crystallization process has been extensively studied [15]. Sonawane et al. [16] synthesized nanosized $CaCO_3$ particles using a sonochemical carbonation process in which CO_2 gas was passed through a hole of an ultrasonic horn to achieve effective micromixing during the synthesis.

To the authors' knowledge, no systematic study has been carried out on the hydrodynamic cavitation-assisted synthesis of $CaCO_3$. Hence, the aims of the current investigation were to synthesize $CaCO_3$ nanoparticles using hydrodynamic cavitation under various experimental conditions and to study the effect of parameters such as reactor design on the crystal size and size distribution.

EXPERIMENTAL

Experimental Apparatus and Procedure

The experimental assembly (Figure 1(a)) consisted of a closed loop reactor with a holding tank of 5 L volume, a centrifugal pump (2880 RPM, 0.5 HP/ 0.37 KW) to recycle the $Ca(OH)_2$ slurry, and two flanges for holding the orifice plate. Downstream pipe (20 mm dia) was made up of a transparent acrylic material in order to observe the generation of cavities. Upstream pipe was made of (SS 316) stainless steel. Pressure gauges were provided to measure the inlet pressure (P_1) and fully recovered pressure (P_2). Three different orifice plates and one having different geometry were used in the present study (Figure 1(b)). CO_2 (99.9% pure) gas was introduced at different flow rates from a hole drilled at the downstream part of pipe. Samples were withdrawn after every 3 minutes interval and were titrated against HCl and the pH and conductivity were measured. $Ca(OH)_2$ of analytical grade was procured and used. HPLC grade water (Millipore) was used for preparing all the slurry suspension and filtered to get uniform particle size of $Ca(OH)_2$. The recycling of slurry enabled the particles to be in suspension without settling at the bottom. The diameter of the orifice was calculated using C_v values which was calculated from P_2 (downstream pressure),   (density of water), V_0 (average velocity near orifice), and P_v (vapor pressure of water) [17].

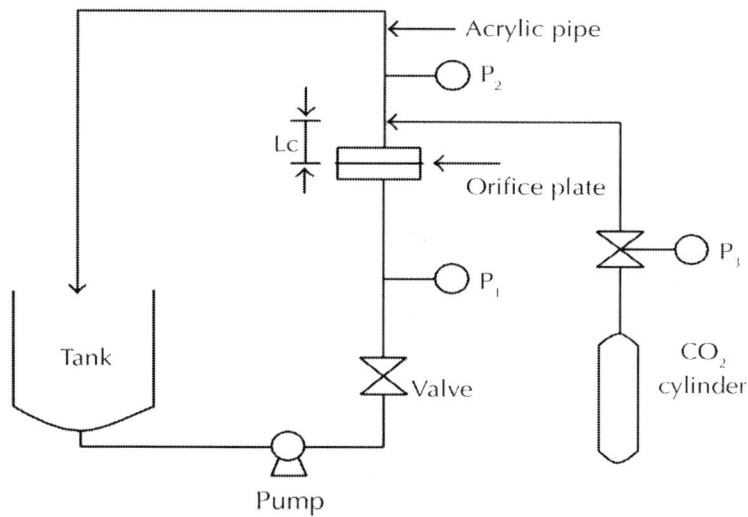

Lc cavitation length P_1, P_2, P_3 pressure gauges

(a)

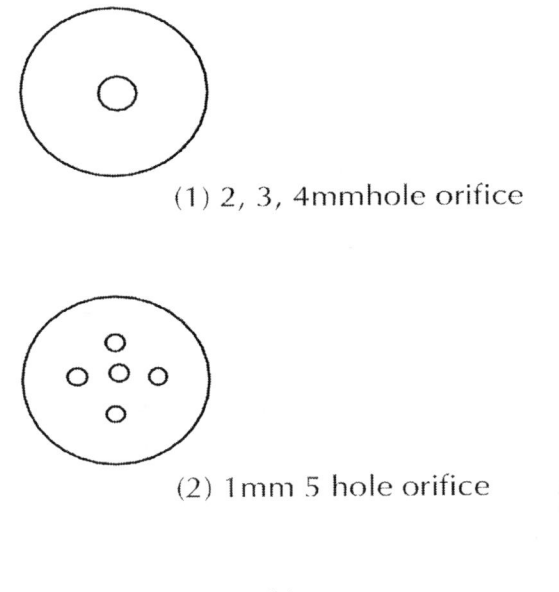

(1) 2, 3, 4mmhole orifice

(2) 1mm 5 hole orifice

(b)

Figure 1: (a) Hydrodynamic cavitation setup for nano-CaCO$_3$ production. (b) Orifice Configurations used for Cavitation.

Structural Properties and Particle Size Characterization

Structural properties of the CaCO$_3$ synthesized by the hydrodynamic cavitation process were evaluated using powder XRD (Philips PW 1800). The Cu-K radiation (LFF tube 35 kV, 50 mA) was selected for the analysis. Transmission electron microscopy (TEM) was performed on Technai G20-stwin working at 200 kV. The measurement of particle size distribution of the nanocrystals was done by dynamic light scattering technique (via Laser input energy of 632 nm).

Cavitation Number (C$_v$)

In order to quantify various cavitation conditions and to represent appropriately the intensity of cavitation, the concept of cavitation number was utilized [6, 17].

The cavitation number is a dimensionless quantity defined as

$$C_v = \frac{(P_2 - P_v)}{(0.5 \times \rho \times V_0^2)},$$

(1)

where C_v is the cavitation number; P_2 is the downstream pressure; P_v is the vapor pressure of water; is the density of water at 25°C, V_0 is the average velocity near orifice.

For different orifice sizes C_v values were calculated using (1) and presented in Table 1. Using appropriate orifice diameter and flow rate to reduce the pressure at vena contracta to fall below the threshold pressure for cavitation (which in many cases is the vapor pressure of water at that temperature) millions of cavities could be produced. The collapsing noise of cavities was heard using stethoscope in all the experiments when an orifice was used.

Table 1: Hydrodynamic conditions of different orifice plate used for experiments

Sr. No	Diameter of orifice	value (ratio of orifice diameter to pipe diameter)	CV(cavitation number)
1	2 mm	0.10	0.15
2	3 mm	0.15	0.75
3	4 mm	0.20	2.15
4	5 holes of 1 mm diameter	0.05	0.21

RESULTS AND DISCUSSIONS

Prior to incorporating orifice into the experimental assembly, it was necessary to carry out a set of experiments to find out the consumption of $Ca(OH)_2$ and the corresponding changes in the solution pH and conductivity (Figure 2). The above three variables were measured for

4% $Ca(OH)_2$ and 5 l/min flow rate of CO_2. All three plots (Figures 2(a), 2(b), and 2(c)) show three distinct regions corresponding to an induction period, nucleation, and precipitation. In the first region, nuclei are formed in the reaction mixture. In the second region nucleation around the crystals takes place, in the third region precipitation process occurs. In case of $Ca(OH)_2$ consumption (Figure 2(a)), first constant period is observed as the solution is saturated with $Ca(OH)_2$. The initial induction period (Figure 2(a)) where no change in $[Ca^{2+}]$ was observed indicates that there was no spontaneous nucleation when the reaction was carried out without the orifice. The nucleation process appears to be slow and it takes relatively longer time (about 15 minutes) for the reaction to complete when the reaction was carried out in the pipe without orifice. It can also be noticed that the carbonation process was not 100% complete and there was some residual $Ca(OH)_2$ present at the end of the reaction. It was also observed that the particles synthesized without the orifice are coarse in nature and they tend to agglomerate during drying.

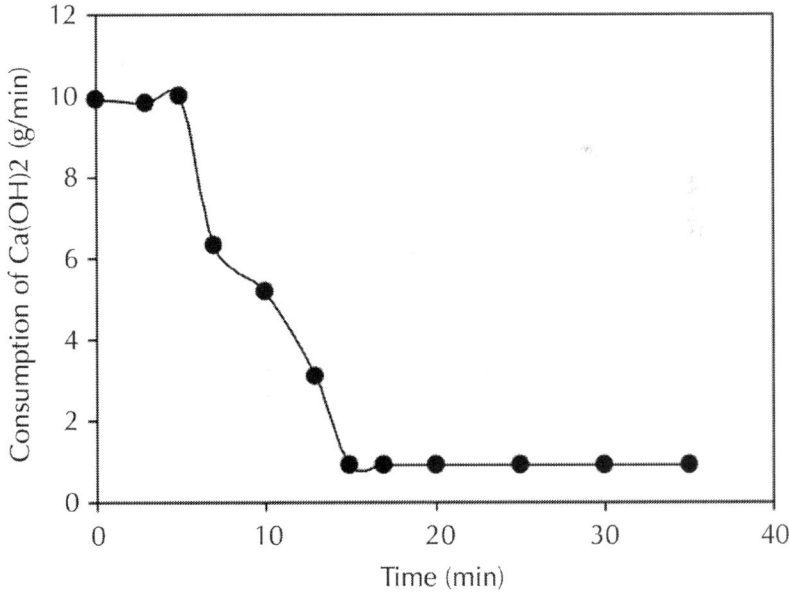

(a)

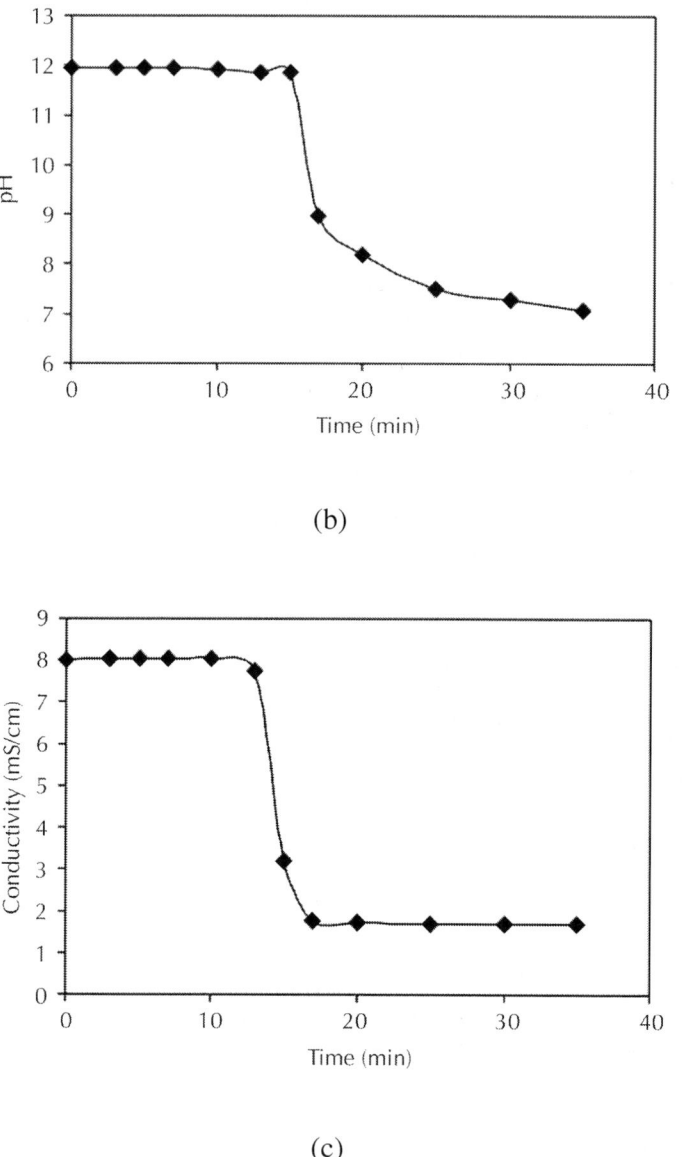

(b)

(c)

Figure 2: (a) Consumption of Ca(OH)$_2$ during the carbonation process without orifice (4% Ca(OH)$_2$ and 5 l/min CO$_2$ flow rate). (b) Change in pH during the carbonation process without orifice (4% Ca(OH)$_2$ and 5 l/min CO$_2$ flow rate). (c) Change in conductivity during the carbonation process without orifice (4% Ca(OH)$_2$ and 5 l/min CO$_2$ flow rate).

It is observed that without the orifice there is a delay in the pH drop (Figure 2(b)) compared to the [Ca^{2+}] drop. This behavior is expected since the carbon dioxide was fed with a constant flow rate with a high initial concentration of $Ca(OH)_2$ which led to no change in pH. This data is also supported by the conductivity measurement (Figure 2(c)), which is an indirect measure of the total concentration of the charged species in the liquid phase. The first region of the conductivity plot gives information about the "induction time", in which the nuclei are formed by absorption of CO_2 onto the surface of $Ca(OH)_2$. The formation of $CaCO_3$ leads to a decrease in the concentration of dissolved ionic species and hence the drop in the conductivity. The relatively slow precipitation process in the absence of the orifice increases the size (101 nm) of the crystals generated, as shown in Table 2 and discussed in the next section.

Table 2: Effect of slurry concentration and CO_2 flow rates, orifice diameter on crystallite size (calculated from XRD), and particle size distribution (from TEM)

Ca(OH)2 Slurry%	CO2 l/ min	Orifice Dia. mm	Crystallite size nm	Particle size distribution nm
Effect of change in CO2 flow rate				
4	3	4	74	65–92
4	5	4	54	62–53
4	7	4	47	35–55
Effect of change in Ca(OH)2 concentration				
2	5	4	50	69–52
4	5	4	54	62–53
6	5	4	61	50–72
Effect of change in orifice diameter				
4	5	1×5 holes	37	29–38
4	5	2	39	30–41
4	5	3	49	43–56
Without orifice				
4	5	4	54	62–53
4	5	—	101	90–168

Effect of Orifice Diameter and Geometry of the Orifice on the Rate of Consumption of Ca(OH)$_2$

Three different orifices with one hole of 2, 3, and 4 mm diameter and one orifice plate having 5 holes of 1 mm diameter were used for cavitation purposes as shown in Figure 1(b). Cavitation number (C_v) and (ratio of orifice diameter to pipe diameter) for the above orifices are presented in Table 1. The consumption profiles of Ca(OH)$_2$ slurry with different orifice diameters are shown in Figure 3. It can be seen that the rate of reaction is much faster with all orifices compared to the rate observed in the absence of an orifice (Figure 2). During the precipitation process, the rate of nucleation step is considered as the rate limiting step. It is observed that the nucleation is enhanced by the hydrodynamic cavitation and hence enhances the rate of precipitation as reported in earlier section. Nishida [18] has observed that acoustic cavitation generated physical effects of microstreaming that influenced the precipitation of calcium carbonate. Our investigation predicts the same effect on the calcium carbonate precipitation by hydrodynamic cavitation. Lyczko et al. [19] reported the effect of cavitation on the primary nucleation of potassium sulphate. They found that the induction period is drastically reduced in the presence of acoustic cavitation. Similar reduction in induction period and enhancement of primary nucleation (no external addition of seed particles) is observed in our case for calcium carbonate precipitation due to hydrodynamic cavitation. As in the case of without orifice, there is an induction period prior to the consumption of Ca(OH)$_2$. However, the length of the induction period is shorter for the 4 mm dia orifice compared to the rest. This indicates that the 4 mm orifice initiates the reaction due to a relatively stronger shear forces generated by the hydrodynamic cavitation process. The larger diameter orifice might have generated the stronger hydrodynamic forces. On changing the geometry of orifice to 1 mm size, 5 holes (Figure 1(b)) orifice geometry, the reaction is initiated without any induction period. This arrangement also yielded smaller particle size during synthesis of calcium carbonate in the carbonation process. X-ray diffraction analysis was carried out for all the powder samples as shown in Figure 4. The grain size (crystallite size) was calculated for all the samples by using Debye Scherrer formula:

$$X_d = \frac{k\lambda}{\beta \cos \theta},$$

(2)

Where k=0.9, = FWHM, and is glancing angle of X-rays with the sample holder, λ =0.5405Å

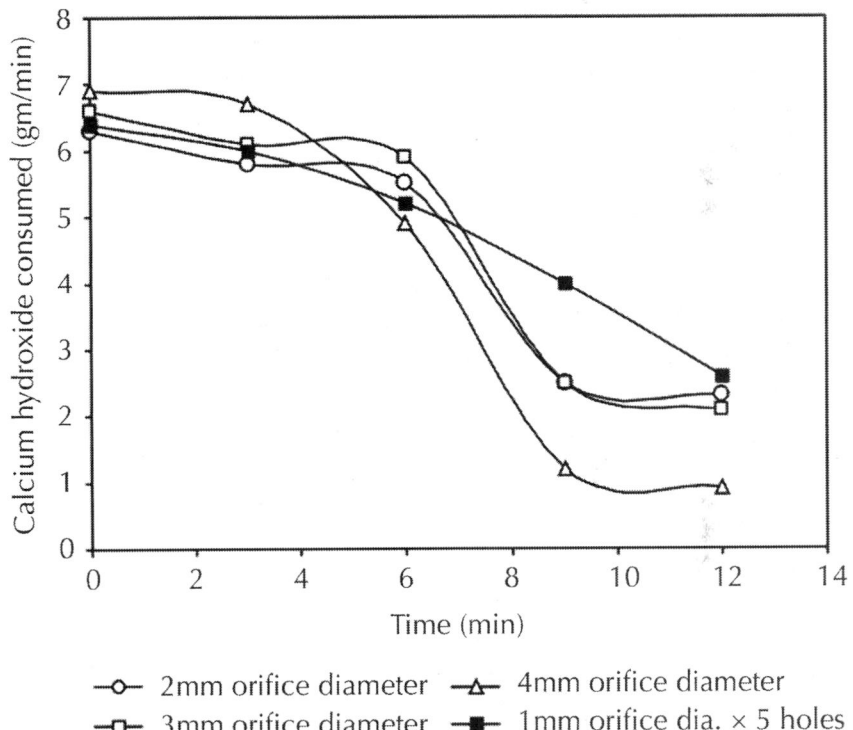

Figure 3: Effect of different orifice on consumption of $Ca(OH)_2$ Slurry with respect of time (showing constant rate period and falling Rate periods) (4% $Ca(OH)_2$ and 5 l/min Co_2 flow rate).

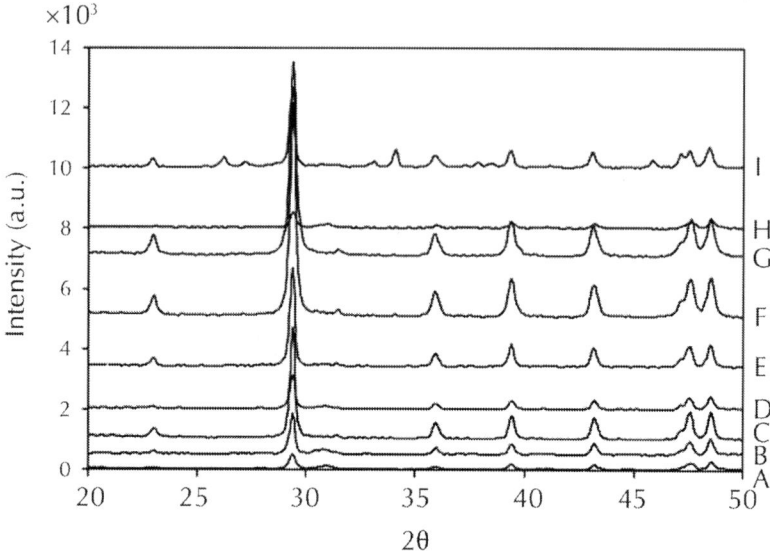

Figure 4: Effect of different experimental conditions on X-ray diffraction spectra. A: Without orifice, 4% Ca(OH)$_2$ slurry, 5 l/min CO$_2$ flow rate; B: 1 mm orifice, 4% Ca(OH)$_2$, slurry 5 l/min CO$_2$ flow rate; C: 2 mm orifice 4% Ca(OH)$_2$, slurry 5 l/min CO$_2$ flow rate; D: 3 mm orifice 4% Ca(OH)$_2$ slurry 5 l/min CO$_2$ flow rate; E: 4 mm orifice 4% Ca(OH)$_2$, slurry 7 l/min CO$_2$ flow rate; F: 4 mm orifice 2% Ca(OH)$_2$, slurry 5 l/min CO$_2$ flow rate; G: 4 mm orifice 6% Ca(OH)$_2$ slurry, 5 l/min CO$_2$ flow rate; H: 4 mm orifice 4% Ca(OH)$_2$slurry, 3 CO$_2$ l/min flow rate; I: 4 mm orifice, 4% Ca(OH)$_2$ slurry, 5 CO$_2$ l/min flow rate.

As shown in Table 2, the 1 mm × 5 holes orifice generated the smallest crystal size of 37 nm. The powder XRD is found to be preferably oriented along (1 0 1 0) plane for samples including the CaCO$_3$ without the orifice. No effect is observed on phases due to change in orifice diameter and geometry of orifice. Single calcite phase is observed in all the samples. Wide particle size distribution was observed for the sample without orifice ranging from 90 to 168 nm as shown in Figure 5(a), while narrow particle size distribution is observed ranging from 30 to 41 nm for the orifice with 2 nm orifice as shown in Figure 5(b). The confirmation of calcite phase observed in the XRD data is confirmed by the TEM images. As shown in Figure 6, the TEM image shows the cubic shape particles of calcite particles synthesized using 4 mm orifice diameter 4% slurry concentration and 5 l/min CO$_2$ flow rate.

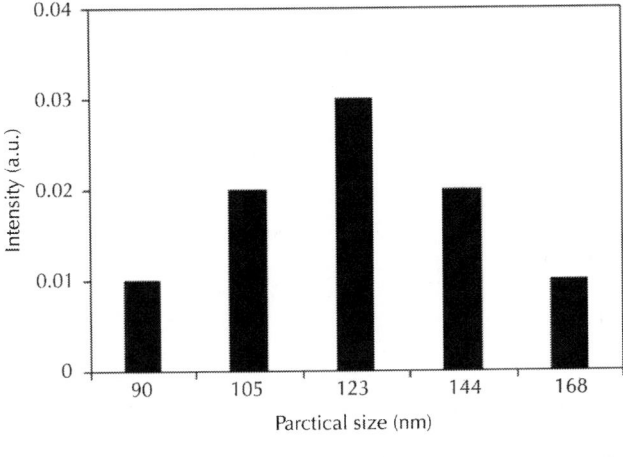

(a)

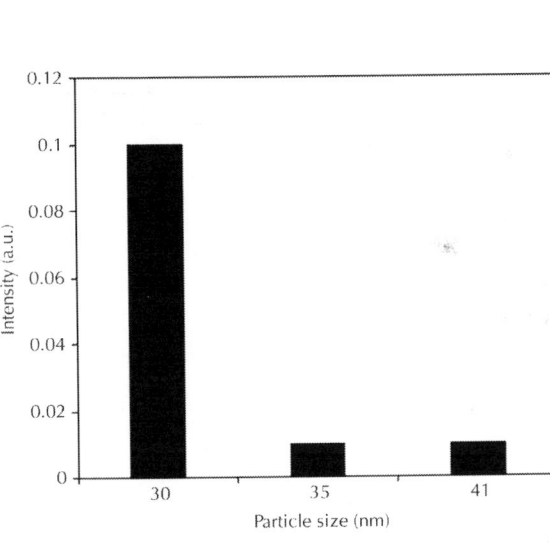

(b)

Figure 5: (a) Particle size distribution of $CaCO_3$ particles synthesized without orifice in the pipe (4% $Ca(OH)_2$ slurry, 5 l/min CO_2 flow rate). (b) Particle size distribution of $CaCO_3$ particles synthesized using Hydrodynamic cavitation (2 mm orifice, 4% $Ca(OH)_2$slurry, 5 l/min CO_2 flow rate).

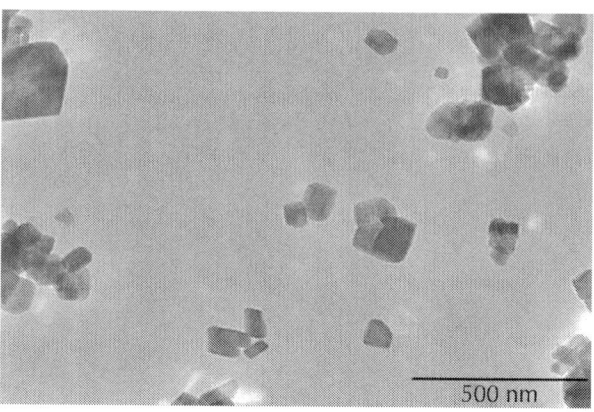

Figure 6: TEM Image of nano-$CaCO_3$ synthesized using hydrodynamic cavitation (4 mm orifice, 4% $Ca(OH)_2$ slurry, 5 l/min, CO_2 flow rate) (Scale 500 nm).

The data shown in Figures 4–6 and Table 2 indicates that the rate of consumption of calcium hydroxide slurry during $CaCO_3$ synthesis using hydrodynamic cavitation is significantly dependent on the geometry of the orifice and its cavitation number. The use of orifice geometry significantly reduces the grain size and hence, increasing the surface area of the synthesized calcite and making it more suitable for the surface coatings applications. Patil and Pandit [7] also have reported the reduction in particle size of styrene butadiene rubber nanoparticles by using the different geometries of orifice used in hydrodynamic cavitation. Luque de Castro and Priego-Capote [20] reported the influence of ultrasonic power and horn tip onto the carbonation process. They found that acoustic cavitation had an effect on the consumption of calcium hydroxide, which shows similar reduction in the induction period and particle size reported in Figure 3 and Table 2, respectively.

Figure 7 shows the effect of cavitation number (C_V) and the ratio of orifice diameter to pipe diameter (β factor) on the rate of consumption of $Ca(OH)_2$ slurry. It is found that the rate of consumption of $Ca(OH)_2$ increases with an increase in the ratio of orifice diameter to plate diameter (β). $Ca(OH)_2$ consumption rate has gone up exponentially with an increase of β factor which indicates that there is a significant effect of geometry of orifice plate. We speculate that the β factor has significant influence on the generation of hydrodynamic cavitation. Generally for effective cavitation, it is recommended that the cavitation number should be less than 2.5 and it is important to know that severity of cavitation increases with decreasing the cavitation number [10], which is the case for all orifices. The higher cavitation number for 4 mm orifice also supports the higher rate observed in Figure 3.

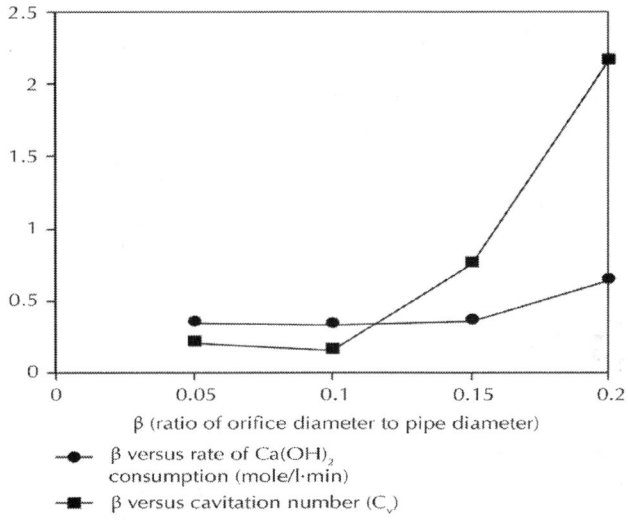

Figure 7: Effect of β factor onto rate of calcium hydroxide consumption and cavitation number (C_v).

Effect of CO_2 Flow Rates onto the Conductivity for Orifice with 4 mm Diameter

A change in the electrical conductivity of the solution with respect to time in the presence of 4 mm diameter orifice is presented in Figure 8. Three different flow rates (3, 5, and 7 l/min) were used at constant 4% $Ca(OH)_2$ concentration. It is found at higher flow rate (7 l/min) that the drop in the solution conductivity is faster in comparison to the other two flow rates. It can also be observed that there is small induction time for higher flow rate 7 l/min while longer induction time for 3 and 5 l/min flow rates. Induction time for 7 l/min flow rate is less than 5 minute, while for 3 and 5 l/min induction time is nearly 10 minute. In comparison to other two flow rates, numbers of nuclei are formed at 7 l/min and hence the drop in conductivity is faster. With increasing the CO_2 flow rate, the reduction in crystallite size and no change in calcite phase are observed as shown in Figure 4 and Table 2. At 3, 5, 7 l/min flow rates, the crystallite size was found 74, 54, and 47 nm, respectively. The particle size distribution is found to be narrow at higher flow rate (7 l/min) ranging from 35 to 55 nm.

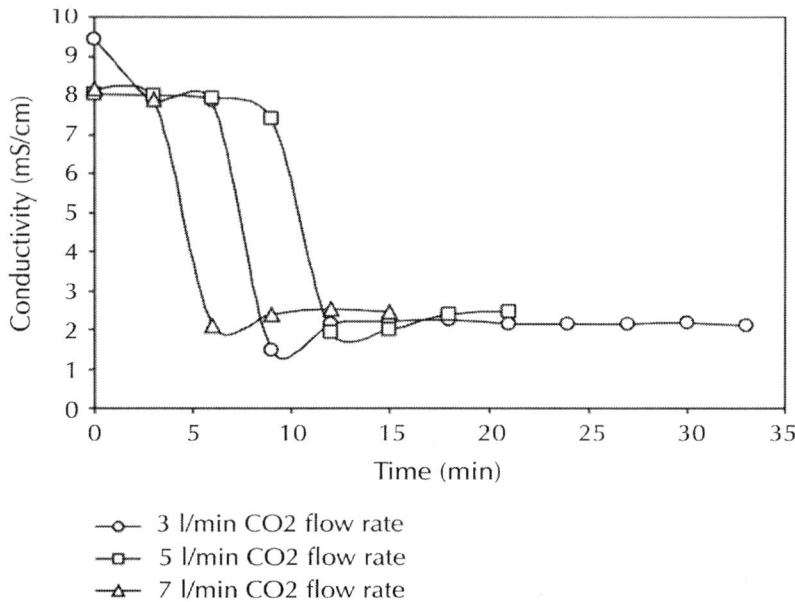

- ─○─ 3 l/min CO2 flow rate
- ─□─ 5 l/min CO2 flow rate
- ─△─ 7 l/min CO2 flow rate

Figure 8: Effect of CO_2 flow rates on conductivity of reaction mixture (orifice diameter 4 mm, 4% $Ca(OH)_2$ slurry concentration).

Effect of $Ca(OH)_2$ Slurry Concentration on the Rate of Reaction

Effect of three different $Ca(OH)_2$ slurry concentrations on the solution conductivity at 4 mm orifice diameter is shown in Figure 9. At higher $Ca(OH)_2$ concentration (6%), the drop in the conductivity takes longer time in comparison to 2% $Ca(OH)_2$ slurry concentration which indicates that there is longer induction time for higher $Ca(OH)_2$ and shorter induction time for low $Ca(OH)_2$ slurry concentration. At 2% of slurry concentration the drop in conductivity occurs very fast (within 5 minutes). It is also found that the available CO_2 gas concentration is more for the 2% slurry and hence induces the massive nuclei formation.

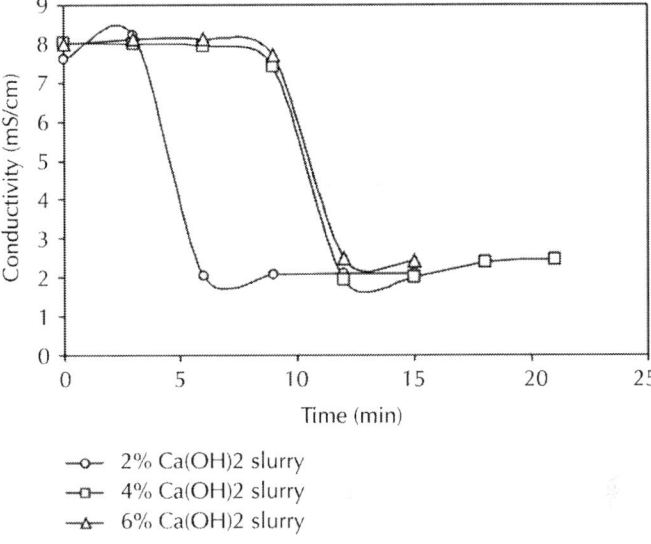

Figure 9: Effect of $Ca(OH)_2$ slurry concentration onto the conductivity at 5 l/min CO_2 gas flow rate.

The effect of calcium hydroxide slurry concentration on the pH values, that is, the completion of reaction was also studied. The reaction was carried out at constant flow rate of CO_2 (5 l/min CO_2 flow rate) and by using 4 mm diameter of orifice. It is seen that the completion of reaction takes longer time (20 minutes) with 6% slurry concentration and on the other hand 2% slurry takes less that 10 minutes for completion. The constant period zone and falling rate zone are observed in the case of 4 and 6% slurry, while 2% slurry shows only falling rate period. As shown in Table 2, with increasing the concentration of calcium hydroxide there is an increase in the crystallite size. This observation suggests that increasing the $Ca(OH)_2$ has no favorable effect on the crystallite size. The XRD patterns for calcite synthesized at different $Ca(OH)_2$ concentration indicate that the calcite powder is crystalline in nature and exhibits pure calcite phase [21].

CONCLUSIONS

This study has demonstrated that nanocalcite can be synthesized by using a hydrodynamic cavitation reactor, without formation of vaterite phase. The effects of three different orifice diameters and the geometry of orifice were evaluated. The average grain size of the calcite synthesized without cavitation was found to be 101 nm. It is seen that there is a wide distribution of particle size in the range 90–168 nm for the setup without an orifice. The change in the geometry of orifice has resulted significant effect on the crystallite size. In case of 5 holes of 1 mm size orifice found to give reduced size of size of 37 nm.

ACKNOWLEDGMENTS

S. H. Sonawane acknowledges the Department of Science and Technology (Govt of India) for providing the funding for fast track project under the Grant no. SR/FTP/ETA-35/2007 and BOYSCAST fellowship through Grant no. SR/BY/E-07/2008. The authors are also thankful to Professor A. B. Pandit (UICT, Mumbai) for his valuable suggestion to carry experimental work.

REFERENCES

1.   A. Gedanken, "Using sonochemistry for the fabrication of nanomaterials," Ultrasonics Sonochemistry, vol. 11, no. 2, pp. 47–55, 2004.

2.   P. R. Gogate, "Cavitational reactors for process intensification of chemical processing applications: a critical review," Chemical Engineering and Processing, vol. 47, no. 4, pp. 515–527, 2008.

3.   P. R. Gogate and A. B. Pandit, "Hydrodynamic cavitation reactors: a state of the art review," Reviews in Chemical Engineering, vol. 17, no. 1, pp. 1–85, 2001.

4.   S. Kakaraniya, A. Gupta, and A. Mehra, "Reactive precipitation in gas-slurry systems: the CO_2-$Ca(OH)_2$-$CaCO_3$ system," Industrial and Engineering Chemistry Research, vol. 46, no. 10, pp. 3170–3179, 2007.

5. A. B. Pandit and J. B. Joshi, "hydrolysis of fatty oils: effect of cavitation," Chemical Engineering Science, vol. 48, no. 19, pp. 3440–3442, 1993.

6. M. M. Chivate and A. B. Pandit, "Effect of sonic and Hydrodynamic cavitation on aqueous polymeric solutions," Indian Chemical Engineering, vol. 35, pp. 52–57, 1993.

7. M. N. Patil and A. B. Pandit, "Cavitation—a novel technique for making stable nano-suspensions,"Ultrasonics Sonochemistry, vol. 14, no. 5, pp. 519–530, 2007.

8. K. R. Morison and C. A. Hutchinson, "Limiting of the Weissler reaction as model reaction for measuring the efficiency of hydrodynamic cavitation," Ultrasonic Sonochemistry, vol. 16, pp. 176–183, 2009.

9. P. Senthil Kumar, M. Sivakumar, and A. B. Pandit, "Experimental quantification of chemical effects of hydrodynamic cavitation," Chemical Engineering Science, vol. 55, no. 9, pp. 1633–1639, 2000.

10. V. S. Moholkar, P. Senthilkumar, and A. B. Pandit, "Hydrodynamic cavitation for sono-chemical effect,"Ultrasonics Sonochemistry, vol. 6, pp. 53–65, 1999.

11. P. R. Gogate and A. B. Pandit, "A review and assessment of hydrodynamic cavitation as a technology for the future," Ultrasonics Sonochemistry, vol. 12, no. 1-2, pp. 21–27, 2005.

12. K. S. Suslick, M. M. Mdleleni, and J. T. Ries, "Chemistry induced by hydrodynamic cavitation," Journal of the American Chemical Society, vol. 119, no. 39, pp. 9303–9304, 1997.

13. J. W. Find and R. Moser, "Preparation and structural properties of Cu-Zn-Al-oxides: a comparative study between the hydrodynamic-cavitation and classical route," Journal of Materials Science, vol. 38, pp. 1917–1924, 2003.

14. R. Lin, J. Zhang, and Y. Bai, "Mass transfer of reactive crystallization in synthesizing calcite nanocrystal,"Chemical Engineering Science, vol. 61, no. 21, pp. 7019–7028, 2006.

15. M. He, E. Forssberg, Y. Wang, and Y. Han, "Ultrasonication-assisted synthesis of calcium carbonate nanoparticles," Chemical Engineering Communications, vol. 192, no. 10–12, pp. 1468–1481, 2005.

16. S. H. Sonawane, S. R. Shirsath, P. K. Khanna, et al., "An innovative method for effective micro-mixing ofCO_2 gas during synthesis of nano-calcite crystal using sonochemical carbonization," Chemical Engineering Journal, vol. 143, no. 1–3, pp. 308–313, 2008.

17. C. Mishra and Y. Peles, "An experimental investigation of hydrodynamic cavitation in micro-Venturis,"Physics of Fluids, vol. 18, no. 10, pp. 103–109, 2006.

18. I. Nishida, "Precipitation of calcium carbonate by ultrasonic irradiation," Ultrasonics Sonochemistry, vol. 11, no. 6, pp. 423–428, 2004.

19. N. Lyczko, F. Espitalier, O. Louisnard, and J. Schwartzentruber, "Effect of ultrasound on the induction time and the metastable zone widths of potassium sulphate," Chemical Engineering Journal, vol. 86, no. 3, pp. 233–241, 2002.

20. M. D. Luque de Castro and F. Priego-Capote, "Ultrasound-assisted crystallization (sonocrystallization),"Ultrasonics Sonochemistry, vol. 14, no. 6, pp. 717–724, 2007.

21. G. Montes-Hernandez, F. Renard, N. Geoffroy, L. Charlet, and J. Pironon, "Calcite precipitation fromCO_2-H_2O-$Ca(OH)_2$ slurry under high pressure of CO_2," Journal of Crystal Growth, vol. 308, pp. 228–236, 2007.

Chapter 5

Influence of the Chemical Composition of Completion Fluids on the Propagation of Electromagnetic Waves within Oil Wells

Alexandre Ashade Lassance Cunha, Marco Aurélio Pacheco, and José Ricardo Bergmann

Departamento de Engenharia Elétrica, Pontifícia Universidade Católica do Rio de Janeiro, Rio de Janeiro, Brazil

ABSTRACT

The propagation of electromagnetic waves in the annular region of oil wells was studied. The present study aims to analyse the propagation attenuation along the well, as well as the input impedance determined by a source placed near the wellhead. A coaxial waveguide model was adopted with heterogeneous dielectrics and losses. First, a wave equation solution for the waveguide is presented, assuming a homogeneous medium with losses, by solving the equation in cylindrical coordinates using the vector potential technique. An uncertainty analysis model is then developed to model the heterogeneous characteristics of the medium. Monte Carlo simulations were performed with the created model using data gathered from the literature. The results of the simulations indicate that propagation in the transverse electromagnetic mode has the smallest attenuation and that for depths of up to 4000 m, there is an attenuation of less than 52 dB. Furthermore, the input impedance ranges from 10 Ω to 10 kΩ because of the uncertainties involved in the problem in question.

INTRODUCTION

Oil wells today are extremely complex and demand very expensive maintenance [1]. Modern drilling techniques can reach a few kilometres in depth, and the costs of a floating platform at open sea can cost billions of Brazilian Reais. Thus, it is critical to determine the internal conditions of the well, and indicators such as temperature, pressure and salinity must be constantly monitored [2].

Reliable telemetry by cables is very difficult to obtain due to the extreme conditions of the internal environment of an oil extraction well. As an example, there is continuous abrasion by sand and dirt that are carried by fluid flow. For this reason, the cables are periodically damaged and require replacements. Such replacements hinder the oil extraction process and, moreover, increase the costs due to the large amount of cables needed over the life of a single well.

The most obvious alternative to cabling is the adoption of wireless telemetry. However, the amount of electricity required to power wireless communication is not practical: power cables are needed because the use of batteries would lead to frequent maintenance, as they need to be recharged [1]. It is obvious, therefore, that the objective (and greater challenge) is to build a system without wires or batteries; that is, the downhole sensors must be powered and communicate with the base without the use of cables.

Several approaches are possible for wireless telemetry. One possibility would be the use of signal transmission techniques using a magnetic field [3]. In this method, a coil that is capable of inducing alternating current through the production pipe is used. This coil has sufficient intensity to transmit both power and the signal itself between the sensor and the base [4]. Through a second coil positioned at the sensor location, it is possible to recover the signal and power needed to feed and produce bilateral communication. However, the technique has a weakness: the production tube is not a transformer core, i.e., it is not designed to minimise the magnetic flux losses. Its hollow characteristic confers much loss by parasitic Foucault currents concomitantly with hysteresis losses, the cause of which is due to the inadequacy of the production tube material for magnetic purposes [5]. Therefore, for long distances, on the order of 3000 m or more, the method is unfeasible.

Another possible approach would be to analyse the oil well from the perspective of a coaxial propagation structure, which is formed by a conductor tube with a perfect centre (production tube) and a cylindrical shell that is coaxial with the tube and is also a perfect conductor. This approach makes it possible to conduct the analysis as it would be performed on a coaxial cable, and therefore, propagation occurs in the transverse electromagnetic mode [6]. This approach, however, has been studied without considering fluids of significant conductivity or possible uncertainties caused by temperature and electrical parameter variations of the fluid that fills the well.

The modelling difficulty concerns the fluid that fills the well. This fluid is heterogeneous because it exhibits significant variation in electric permittivity along the well depth. Furthermore, the actual fluid concentration varies from well to well, and at times within the well itself, which is another source of heterogeneity.

In this context, the present article proposes to study the electromagnetic propagation inside the annular area of oil wells. The goal is to develop a model that permits an approximation of the behaviour of the TEM propagation mode in the coaxial structure with losses and the previously cited heterogeneous characteristics. The first step is the deduction of the electric and magnetic field equations within the well, assuming a typical well structure. Thereafter, statistical models of the parameters that compose the medium are developed, considering their temperature variations and propagation frequency. Finally, the study uses Monte Carlo analysis to illustrate how the propagation attenuation behaves and to analyse the input impedance of the coaxial waveguide as a function of position.

MODELLING

Well Definition

The well will be modelled according to Figure 1. This figure depicts a coaxial guide bounded by perfect conductor metals with a homogeneous dielectric and losses. The upper extremity is completely closed by a perfect conductor metal to model the valve and metal duct assembly present in a wellhead. The bottom of the well is composed of concrete; however, this fact is neglected, and the well is assumed to behave as a semi-infinite waveguide.

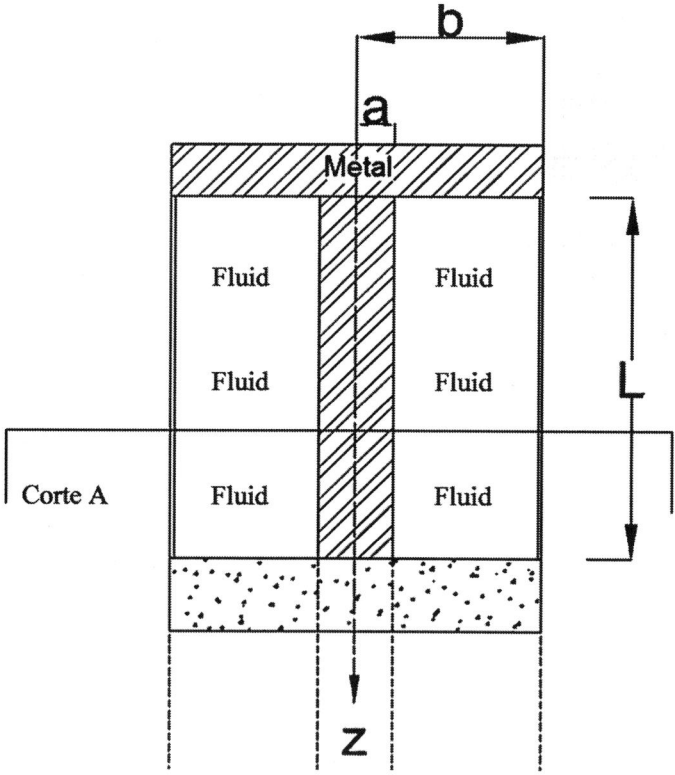

Figure 1: Adopted well structure. A semi-infinite well is assumed and is formed by two concentric cylinders and a metal block covering the top of the waveguide.

Deduction of the Propagation Equations

Modelling is performed using Maxwell's equations and the concept of vector potentials. The Maxwell's equations used are Equations (1) and (2):

$$\nabla \times E = -M_i - j\omega B \tag{1}$$

$$\nabla \times H = J_i + \sigma_e E + j\omega\varepsilon' E \tag{2}$$

The real number constant ε' is the real part of the complex electrical permittivity of the medium. Furthermore, the complex constant $_e$ is called effective conductivity and it represents the linear relation of conduction current and electric field on the medium.

Consider a medium with no source of magnetic charge ($\nabla \cdot B = 0$). Because $\nabla \times \tilde{N} \times = 0$, the curl of A can be defined as

$$\nabla \times A \equiv B = \mu H \tag{3}$$

Substituting 3 in 1 and considering $M_i = 0$ (Induced magnetic flux equal to zero), we obtain:

$$\nabla \times E = -\nabla \times j\omega A \tag{4}$$

Or:

$$\nabla \times (E + j\omega A) = 0 \tag{5}$$

Since $\tilde{N}'\nabla = 0$, an electric scalar potential g_e can be defined such that:

$$E + j\omega A \equiv -\nabla g_e \tag{6}$$

Applying the identity $\tilde{N}'\tilde{N} \times = \nabla(\nabla) - \tilde{N}^2$ in Equation (3) and using Equation (2), we obtain:

$$-\nabla\nabla \cdot A + \nabla^2 A$$

$$= -\mu J_i + \mu\left(\sigma_e + j\omega\varepsilon'\right)\left(\nabla g_e + j\omega A\right)$$

$$\sigma_e \equiv \omega\varepsilon'' + \sigma_s \tag{7}$$

In Equation (7), the parameter σ_e is the effective conductivity of the medium, while σ_s represents the static conductivity, that is, the conductivity when frequency is zero. At this moment, we are in a position to define ∇A. To simplify Equation (7), we use the gauge:

$$\nabla \cdot A \equiv -\frac{\gamma^2}{j\omega} g_e \tag{8}$$

Where $\gamma_1^2 \equiv j\omega\mu(\sigma_e + j\omega\varepsilon')$ is the complex constant of propagation of the medium? This step simplifies the equation for the electrical potential, resulting in:

$$\nabla^2 A = -\mu J_i + \gamma^2 A \tag{9}$$

For homogeneous media, Equation (9), when solved, determines the electric vector potential in the medium, which can be used to obtain the equation for the electric field as a function of the electric vector potential:

$$E = \frac{j\omega}{\gamma^2}\nabla\nabla \cdot A - j\omega A \tag{10}$$

We seek the solution in transverse electromagnetic mode (TEM), which is generated using $A=\hat{a}_z A_z$ with the restriction $E_z = 0$, which leads to the following fields in cylindrical coordinates:

$$E_\rho = \frac{j\omega}{\gamma_1}\frac{A}{\rho}\sinh\left(\gamma_1 z\right) \tag{11}$$

$$H_\phi = -\frac{1}{\mu}\frac{A}{\rho}\cosh(\gamma_1 z)$$

(12)

$$\gamma_1^2 \equiv j\omega\mu(\sigma_e + j\omega\varepsilon')$$

(13)

Note that because $E\Phi = E_z = 0$, the boundary conditions that require a null tangential component in the metal extremities are already met. As the well has a metal structure at one of the extremities, one must also ensure that $E\rho$ ($z = 0$) = 0. Note, however, that this condition is also already guaranteed.

The above solution represents a propagation model in TEM mode in a coaxial medium with losses inherent to the medium inside. However, it is important to remember that the solution was obtained by assuming a homogeneous propagation medium.

Statistical Modelling of the Medium Constituent Parameters

The propagation medium in the well consists of an oil based dielectric fluid composed of water, oil and salts (typically $CaCl_2$) [7]. This fluid is the centre of all electromagnetic propagation, and therefore, its electrical characteristics must be examined in detail. A study over the range of 1 MHz to 100 MHz was previously conducted [7], revealing significantly variable behaviour based on the chemical composition of the fluid.

To model the variation of conductivity with frequency, the effective conductivity concept is applied [8] using the following formula:

$$\sigma_e \equiv \sigma_s + \omega\varepsilon''$$

(14)

From the experimental curves obtained in [8], the parameters s_s and ε'' can be calculated using a least squares method.

An analogous model can be created to model the variation in frequency of the real relative permittivity with the frequency. With

$$\varepsilon' \equiv \varepsilon_s + \kappa\omega \tag{15}$$

And using the experimental curves in [8], it is possible to estimate the parameters and κ using a least squares approach.

In addition to the variation in frequency, variation in the medium e_s constituent parameter can also be observed in temperature. Such variation cannot be neglected because inside a 5000 m deep well, it is impossible to ensure that the temperature is uniform along its entire length. This fact, therefore, characterises non-homogeneity along the z direction.

To circumvent the situation, a model that assumes an average temperature in the medium is adopted. This temperature, in turn, is considered constant throughout the well, which implies a homogeneous medium. Thus, the influence of the variation of this average temperature on signal propagation in the well can be analysed, assuming a valid range of average temperatures.

Mathematically, a coefficient of correction in temperature is defined as the ratio between the temperature value in question and the value of the reference temperature, here defined as 25°C. Thus,

$$\theta \equiv \frac{\sigma_e(\theta)}{\sigma_e(25°C)} \tag{16}$$

$$\theta \approx a_1\theta^2 + a_2\theta + a_3 \tag{17}$$

$$\sigma_e = \sigma_e(25°C) \times \theta \tag{18}$$

The quadratic form was selected to present the best interpolating results using the experimental data from [8].

Similarly, there is increasing variation in the relative permittivity according to the temperature, as demonstrated by [7]. Again, it is used a quadratic model similar to the effective conductivity variation in temperature:

$$\theta_\varepsilon \equiv \frac{\varepsilon'(\theta)}{\varepsilon'(25°C)}$$

(19)

$$\theta_\varepsilon \approx a_1\theta^2 + a_2\theta + a_3$$

(20)

$$\varepsilon' = \varepsilon'(25°C) \times \theta_\varepsilon$$

(21)

EXPERIMENTS AND RESULTS

In all the experiments, a well with a length of 5000 m, an inner radius of 0.05 m and an outer radius of 0.1 m is assumed. All the Monte Carlo analyses were conducted with at least 1 million samples. The objective of the experiments was to obtain graphs and numerical values for the input impedance of the "oil well" waveguide and to analyse the propagation loss in the medium for 5000 m of depth.

All analyses were conducted over the range of 1 MHz to 100 MHz. For the other free parameters, the modelling used random variables whose distribution was selected to reflect their most common values. Table 1 summarises the values selected for each free parameter of the previously described model and their respective distributions. For simplicity, when the random variable has simple and obvious domain restrictions, a uniform distribution was selected; otherwise, a normal distribution was used. Furthermore, statistical independence was assumed between the variables.

First, an experiment was conducted to calculate the attenuation in the well. The attenuation is directly dependent on the real part of the constant of propagation, whose square is defined as $\gamma_1^2 \equiv j\omega\mu(\sigma_e + j\omega\varepsilon')$

.

The models proposed in Section 2.3 were used for the conductivity and electric permittivity, and the relative magnetic permeability was assumed to be equal to 1.

The Figure 2 presents a boxplot of the constant of attenuation for the various frequencies between 1 and 100 MHz. Values that appear in the graph as outliers are from highly unlikely combinations of the random variables of the problem and are most likely not physically feasible and, therefore, must be disregarded. The height of the boxes represents the range of values most likely to be observed, and its average point is a good approximation of the average. The boxplot of Figure 3 demonstrates that the attenuation constant increases with an increase in frequency, as expected. Furthermore, note that the deviation of the coefficient increases with the elevation of the propagation frequency, which makes the system design even more difficult. Thus, it is clear that lower frequentcies are desirable from the point of view of signal attenuation.

Table 1: Random variables used for modelling uncertainties. For simplicity, when the random variable has simple and obvious domain restrictions, a uniform distribution was selected; otherwise, a normal distribution was used

Variable	Average/a	Deviation/b	**Distribution**
σ_s (S/m)	0	4.0×10^{-5}	Uniform
ε_s (F/m)	3	12	Uniform
ε' (F/m)	1.0×10^{-13}	3.0×10^{-12}	Uniform
k (F/H$_z$·m)	2.5×10^{-9}	5.0×10^{-10}	Normal
θ_{med} (°C)	40	5	Normal
a_1 (permissivity)	9.0×10^{-4}	2.0×10^{-4}	Normal
a_2 (permissivity)	-3.0×10^{-12}	6.0×10^{-3}	Normal
a_3 (permissivity)	1.2	0.1	Normal
a_1 (permissivity)	1.0×10^{-5}	2.0×10^{-6}	Normal
a_2 (permissivity)	0.0	5.0×10^{-4}	Normal
a_3 (permissivity)	1.0	1.0×10^{-2}	Normal

Using the TEM mode equations presented in Equation (2), it is clear that the wave propagating in the direction of the bottom of the well has power attenuation given by A=exp (a²L²). Using the statistical analysis of the coefficient of attenuation, Table 2 was generated.

The values from the table represent the upper limit of attenuation for 95% of the cases. Thus, for example, for a depth of 1000 m, P (A ≤ 13 dB) = 0.95. The results presented in Table 2 do not agree with the propagation for L = 5000 m of depth because of the high-energy attenuation (130.3 dB). However, it is essential to note that the table represents an upper limit of attenuation, taking 95% of the possible combinations of propagation medium and an average temperature. Moreover, as the table demonstrates, propagations up to 2000 m depth are highly acceptable, as an attenuation of up to 52.1 dB is observed in current communication technology. Even at greater depths, the possibility of communication cannot be excluded.

The second experiment was performed to analyse the input impedance of the coaxial waveguide as a function of the excitation source position. The experiment was conducted at 1 MHz.

Setting the impedance at any point as the ratio E_ρ / H_ϕ, we obtain

$$Z = Z_0 \tanh\left(\gamma_1 z\right)$$
(22)

$$-L < z \leq 0$$
(23)

$$Z_0 = -\mu \frac{j\omega}{\gamma_1}$$
(24)

The position selected for analysis was at 1/4 wavelength from the wellhead, that is, $Z = \lambda/4$ however, the wavelength itself is a random variable because $\lambda = 2*\pi/\beta$ and $\beta = \text{Im}\{g\}$ is a random variable. Monte Carlo analysis revealed that λ varies between 90 m and 170 m for 90% of the cases, leading to the selection of position $Z = ((90+170)*1/2)/4 = 32.5 m$

Table 2: Estimated attenuation of electromagnectic waves propagating in the annular region of the well

L (m)	Electric field attenuation (dB)	Power attenuation (dB)
1000	13.0	26.1
2000	26.1	52.1
3000	39.1	78.2
4000	52.1	104.2
5000	65.1	130.3
6000	78.2	156.3
7000	91.2	182.4

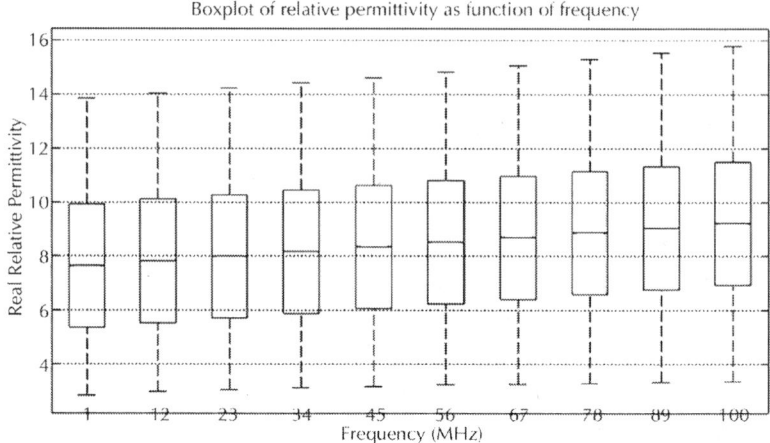

Figure 2: Boxplot of relative permittivity as a function of frequency. The horizontal axis represent the frequency of propagation in MHz, while the vertical axis represent the actual value of the electrical permittivity.

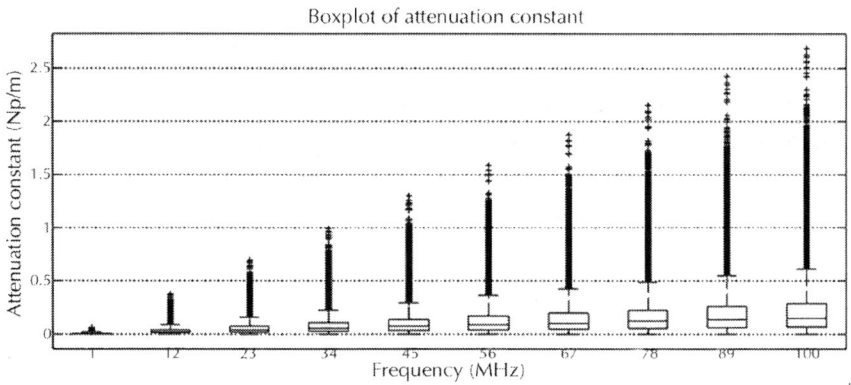

Figure 3: Boxplot of attenuation constant as a function of frequency.

Using Monte Carlo analysis again, it can be observed in Figure 4 that the input resistance (real part of the impedance) of the system exhibits great variation, ranging from 10 Ω to 1.0 kΩ for 90% of the cases. This variation is due to the variation of the relative permittivity in the medium. Therefore, it is necessary to design a generator circuit that provides a good match for a wide range of input impedances, i.e., the generator/receiver circuit must lose as little power as possible by reflection.

CONCLUSIONS

The present study focused on analysis of the electromagnetic propagation inside the annular area of oil wells, assuming the interior medium is composed of dielectrics with significant conductivity. The well behaviour was evaluated with respect to the input impedance and propagation attenuation.

To quantify the well behaviour with respect to electromagnetic propagation, a coaxial waveguide model was developed, modelling the wellhead as a metal cap that completely closes one extremity of the waveguide. First, we solved the wave equation for a homogeneous cylindrical coaxial waveguide, resulting in an analytical model. Then, an uncertainty model was adopted to approximate the heterogeneous and imprecise characteristics of the fluid that functions as a dielectric inside the well.

Using this model, two experiments were developed, both by simulation. The first aimed at analysing the input impedance observed by a source positioned at 1/4 of the wavelength from the wellhead, while the other aimed at analysing the attenuation as a function of length or well depth.

Due to the uncertainties present, the input impedance was observed to vary from 10 Ω to 1 kΩ for the frequency of 1 MHz. The conclusion was obtained by Monte Carlo simulation and was applied to the expression derived for the waveguide input impedance.

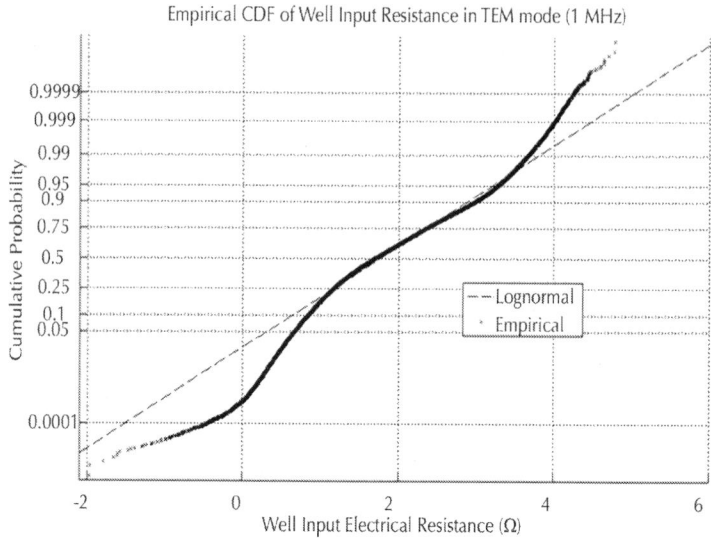

Figure 4: Empirical cumulative distribution function of the well input resistance in TEM mode at 1/4 wavelength from well head. The horizontal axis show the log on base 10 of the value of input resistance in Ohms, while the vertical axis show the cumulative distribution function value.

By performing the attenuation analysis, it was concluded that for 95% of the cases, the constant of attenuation in TEM mode is less than $0.8*10^{-4}$ Np/m for a frequency of 1 MHz. The power attenuation at 4000 m depth was also observed to be approximately 100 dB for the same frequency of 1 MHz, with 95% probability. Although 100 dB appears to be a large attenuation, most wells in operation are between 1000 m and 2000 m in length, and the attenuation is much lower at these depths.

REFERENCES

1.   S. Brilles, "Remote Downhole Well Telemetry," US Patent US6766141, 2004.

2.   J. A. D. Rosa, A. J. Carvalho and R. D. S. Xavier, "Petroleum Reservoir Engineering," Rio de Janeiro, 2006.

3.   F. Sakata, H. Wakiwaka, M. Hanabusa, N. Yamazaki and H. Yamada, "Performance Analysis of Long Distance Transmitting of

a Magnetic Signal in a Cylindrical Steel Rod," IEEE Translation Journal on Magnectics in Japan, Vol. 8, No. 2, 1993, pp. 102-106.

4.   F. Harold J. Vinegar, R. R. Burnett, G. C. W. M. Savage and J. W. Hall, "Permanent Downhole, Wireless, TwoWay Telemetry Backbone Using Redundant Repeaters," US Patent US6633236B2, 2003.

5.   B. W. Kennedy, "Energy Efficient Transformers," New York, 1998.

6.   K. A. Safynia and R. W. McBride, "System and Method for Communicating Signals in a Cased Borehole with Tubing," US Patent US4839644, 1989.

7.   P. A. Patil, et al., "Experimental Study of Electrical Properties of Oil-Based Mud in the Frequency Range from 1 to 100 MHz," SPE Drilling and Completion, Vol. 25, No. 3, 2010, pp. 380-390. doi:10.2118/118802-PA

8.   C. A. Balanis, "Advanced Engineering Electromagnetics, Vol. 52, No. 1," Wiley, Hoboken, 1989, p. 1008.

Effects of Chemical Reaction on the Unsteady Free Convection Flow past an Infinite Vertical Permeable Moving Plate with Variable Temperature

Fayza Mohammed Nasser El-Fayez

Department of Mathematics, College of Science, Princess Norah Bint Abdulrahaman University, Riyadh, Saudi Arabia

ABSTRACT

Analytical solutions for the effect of chemical reaction on the unsteady free convection flow past an infinite vertical permeable moving plate with variable temperature has been studied. The plate is assumed to move with a constant velocity in the direction of fluid flow. The highly nonlinear coupled differential equations governing the boundary layer flow, heat and mass transfer are solved using two-term harmonic and non-harmonic functions. The parameters that arise in the perturbation analysis are Prandtl number (thermal diffusivity), Schmidt number (mass diffusivity), Grashof number (free convection), modified Grashof number, Chemical reaction parameter (rate constant), Skin friction coefficient and Sherwood number (wall mass transfer coefficient). The study has been compared with available exact solution in the literature and they are found to be in good agreement. It is observed that: The concentration increases during generative reaction and decreases in destructive reaction. The concentration increases with decreasing Schmidt number. The effect of increasing values of K leads to a fall in velocity profiles. The velocity decreases with increasing values of the Schmidt number. An increase in modified Grashof number leads to an increase in velocity profiles. The skin friction increases with decreasing Schmidt number. In generative reaction the skin friction decreases and in destructive reaction the skin friction increases.

INTRODUCTION

Chemical heat and mass transfer in natural convection flows on vertical cylinders has a wide range of applications in the field of science and technology [1]. In systems where a chemical reaction of dissociation recombination takes place, the total heat transfer may be increased to the energy transfer by diffusion under the influence of a concentration gradient. Heat transfer and its related topics are heavily studied fields [2]. These studies range from over simplified problems to highly complex types of interactions and configurations which requires sophisticated numerical schemes and high speed computers to obtain reasonably accurate results [3]. However the common ground for most of these studies is that they are solved and analyzed by assuming a pure fluid with no contaminants. While this assumption approximates the reality in some cases quite well, especially for low particle contamination levels, it is not, however, valid in a lot of other cases in which the contaminants in the fluid play a major role in altering the resultant flow and heat transfer characteristics [4]. In many world environments, such as in Saudi, for example, dust storms or fine dust suspension in the air we encountered for many months during the year. These fine particles of dust penetrate the enclosures and the various devices, and have a serious impact on the performance of many equipment.

This example represents a situation of particulate suspension where a pure fluid assumption does not accurately represent the reality. The outlook for a direct coal fired magnetohydrodynamic (MHD) power generator as potentially significant source of energy seems promising in view of its efficiently, its effect on environment and the availability of needed natural resources. These studies are useful in understanding the effect of the presence of a slag layer on heat transfer characteristics of a coalfired magnetohydrodynamic (MHD) generator. A fiew representative fields of interest in which combined heat and mass transfer along with chemical reaction play an important role in chemical process industries such as food processing and polymer production. Bottemanne [1] gave the experimental results for a vertical cylinder with simultaneous heat and mass transfer and evaporation of water vapour into still air with Prandtl number ($P_r = 0.71$)

and Schmidt number ($S_c = 0.63$) under the uniform wall temperature/concentration condition. Bottemanne's [1] experiments were conducted using a cylinder of "large diameter". The effects of mass transfer on the flow past an impulsively started infinite vertical plate under constant heat flux condition along with chemical reaction were studied by Das et al. [2]. Exact solutions were derived by the Laplace-transformation technique. They observed that the skin friction is positive, at large values of the chemical reaction parameter. Sakiadis [3,4] studied the growth of the two dimensional velocity boundary layer over a continuously moving horizontal plate, emerging from a wide slot, at uniform velocity. Soundalgekar [5] has studied mass transfer effects on flow past an impulsively started infinite isothermal vertical plate.

Again, Soundalgekar et al. [6] analyzed the mass transfer effects on impulsively started infinite vertical plate with variable temperature or uniform heat flux. Chambre and Young [7] have studied a first order chemical reaction in the neighborhood of horizontal plate. Ramanamurthy and Govinda Rao [8] presented the results for flow past an impulsively started infinite vertical plate. They gave the detailed effects of mass transfer on the plate along with heat flux and chemical reaction. The effects of mass transfer on the flow past an impulsively started infinite vertical plate under constant heat flux condition were studied by Das et al. [9]. Ganesan and Rani [10] presented a numerical study of transient flow along a semi infinite vertical cylinder subjected to a uniform wall temperature and concentration. Ganesan and Rani [11] carried out a numerical study of transient natural convection flow over a vertical cylinder under the combined buoyancy effects of heat and mass transfer along with chemical reaction. Some works are available in the subject of MHD convection by Abdelkhalek [12-19], Benjamin Gebhart et al. [21] and Muthucumaraswany R. and Ganesan P. [22]. Unsteady MHD convection and mass transfer flow of micropolar fluid past a vertical permeable moving plate were studied by Zarea et al. [20].

It is proposed to study, the flow past an impulsively started infinite vertical plate with variable temperature and uniform mass diffusion in the presence of a homogeneous chemical reaction. The main reason for the lack of study of this problem is due to difficult mathematical and numerical procedures in dealing with the non-similar boundary layers. The highly non-linear coupled differenttial equations governing the boundary layer flow, heat and mass transfer are solved using two-term harmonic and non harmonic functions. Details of the velocities, temperature and concentration fields as well as the local skin friction and the local Sherwood number for the various values of the parameters of the problem are presented.

MATHEMATICAL ANALYSIS

Consider unsteady two-dimensional flow of a laminar, viscous, and heat absorbing fluid past an infinite vertical permeable moving plate. The axial coordinate x' is measured vertically upward along the plate, and the y' axis is taken normal to the plate. At time $t' \leq 0$ the plate and fluid are at the same temperature T'_∞ and concentration C'_∞. At time $t' > 0$, the plate is given an impulsive motion in the vertical direction against the gravitational field with uniform velocity u_0, the plate temperature is made to raise linearly with time. Also the level of the species concentration is raised to C'_W. It is also assumed that there exists a homogeneous first order chemical reaction between the fluid and species concentration. But here we assume the level of species concentration to be very low and hence heat generated during chemical reaction can be neglected. In this reaction the reactive component given off by the surface, occurs only in very dilute form. Hence, any convective mass transport to or from the surface due to a net viscous dissipation effects in the energy equation are assumed to be negligible. Under these assumptions, the boundary layer flow with Boussinesq's approximation is governed by:

$$\frac{\partial u'}{\partial t'} = \nu \frac{\partial^2 u'}{\partial y'^2} + g\beta(T' - T'_\infty) + g\beta^*(C' - C'_\infty) \tag{1}$$

$$\rho C_P \frac{\partial T'}{\partial t'} = K \frac{\partial^2 T'}{\partial y'^2} \tag{2}$$

$$\frac{\partial C'}{\partial t'} = D \frac{\partial^2 C'}{\partial y'^2} - K_l C' \tag{3}$$

where, u' is the velocity of the fluid in the x' direction, g is the acceleration due to gravity,   is the volumetric coefficient of thermal expansion,   * is the volumetric coefficient of expansion with concentration, T' is the temperature of the fluid near the plate, T'_∞ is the temperature of the fluid away from the plate, T'_W is the surface temperature t' time, t is the dimensionless time,   is the kinematic viscosity, C› species concentration, C dimensionless species concentration, C_∞ species concentration away from the plate, C_W the surface species concentration, D mass diffusion coefficient, K thermal conductiveity, K_l chemical reaction parameter,   density of the fluid, C_p specific heat at constant pressure, u dimensionless velocity, y' coordinate axis normal to the plate, y dimensionless coordinate axis normal to the plate.

With the following initial and boundary conditions:

$$u' = 0, \ T' = T'_\infty, \ C' = C'_\infty \text{, for all } y', \ t' \le 0$$

$$t' > 0: \ u' = u_0, \ T' = T'_\infty + (T'_w - T'_\infty) At', \ C' = C'_w$$

at $y' = 0$

$$u' = 0, \ T' \to T'_\infty, \ C' \to C'_\infty \text{ as}$$

$$y' \to \infty$$

(4)

where, $A = u_0^2/v$, u_0 velocity of the plate.

We now introduce the following non-dimensional quantities:

$$u' = u u_0 G, \ t' = \frac{tv}{u_0^2}, \ y' = \frac{yv}{u_0}, \ \theta = \frac{T' - T'_\infty}{T'_w - T'_\infty},$$

$$G = \frac{g\beta v \left(T'_w - T'_\infty\right)}{u_0^3}, \ C = \frac{C' - C'_\infty}{C'_w - C'_\infty},$$

$$G_0 = \frac{v g \beta^* \left(C'_w - C'_\infty\right)}{u_0^3}, \ P_r = \frac{\mu c_P}{K}, \ S_c = \frac{v}{D}, \ K = \frac{v k_l}{u_0^2}$$

(5)

where, G Grashof number, G_0 modified Grashof number, S_c Schmidt number, μ coefficient of viscosity,   dimensionless temperature.

In Equations (1)-(4), which leads to

$$\frac{\partial u}{\partial t} = \theta + \frac{G_0}{G} C + \frac{\partial^2 u}{\partial y^2}$$

(6)

$$\frac{\partial \theta}{\partial t} = P_r^{-1} \frac{\partial^2 \theta}{\partial y^2}$$

(7)

$$\frac{\partial C}{\partial t} = S_c^{-1} \frac{\partial^2 C}{\partial y^2} - KC$$

Therefore,

$$\frac{\partial C}{\partial t} = S_c^{-1}\frac{\partial^2 C}{\partial y^2} - KC$$

(8)

Equations (6)-(8), represent a set of partial differential equations that can not be solved in enclosed form. However, it can reduced to a set of ordinary differential equations in dimensional form that can be solved analytically, this can be done by representing the velocity, temperature and the concentration as:

$$u = u_0 + \varepsilon e^{i\omega t} u_1 + \varepsilon^2 e^{2i\omega t} u_2$$

(9)

$$\theta = \theta_0 + \varepsilon\, e^{i\omega t}\theta_1 + \varepsilon^2\, e^{2i\omega t}\theta_2$$

(10)

$$C = C_0 + \varepsilon\, e^{i\omega t} C_1 + \varepsilon^2 e^{2i\omega t} C_2$$

(11)

Substituting Equations (9)-(11) into Equations (6)-(8), equating the harmonic and non harmonic terms and neglecting the higher order of (^{3}), and simplifying we obtain the following set of differential equations for u, and C.

$$u_0'' = -\theta_0 - \frac{G_0}{G}C_0$$

(12)

$$u_1'' - i\omega u_1 = -\theta_1 - \frac{G_0}{G}C_1$$

(13)

$$u_2'' - 2i\omega u_2 = -\theta_2 - \frac{G_0}{G}C_2$$

(14)

$$P_r^{-1}\theta_0'' = 0$$

(15)

$$\theta_1'' - i\omega P_r\theta_1 = 0$$

(16)

$$\theta_2'' - 2i\omega P_r \theta_2 = 0 \tag{17}$$

$$C_0'' - S_c K C_0 = 0 \tag{18}$$

$$C_1'' - S_c (K + i\omega) C_1 = 0 \tag{19}$$

$$C_2'' - S_c (K + 2i\omega) C_2 = 0 \tag{20}$$

In the above equations, the primes denote differentiation with respect to y.

The boundary conditions (4) after substitution Equations (9)-(11) are reduced toat

$$y = 0, \qquad u_0 = \frac{1}{G}, \ u_1 = 0, \ u_2 = 0$$

$$\theta_0 = 1, \ \theta_1 = 0, \ \theta_2 = 0$$

$$C_0 = 1, \ C_1 = 0, \ C_2 = 0$$

at $y \to \infty$, $\qquad u_0 = 0, \ u_1 = 0, \ u_2 = 0$

$$\theta_0 \to 0, \ \theta_1 \to 0, \ \theta_2 \to 0$$

$$C_0 \to 0, \ C_1 \to 0, \ C_2 \to 0 \tag{21}$$

Hence from Equations (18)-(20) under the respective boundary conditions (21), and substituting the solutions into Equation (11) the solution for concentration distribution is given by:

$$C = e^{-\sqrt{S_c K} y} + \varepsilon e^{i\omega t} e^{-\sqrt{S_c (K+i\omega)} y} + \varepsilon^2 e^{2i\omega t} e^{-\sqrt{S_c (K+2i\omega)} y}$$

Also, by solving the differential Equations (15)-(17), under the boundary conditions (21), and substituting the solutions into Equation (10). We have the temperature distribution is given by:

$$\theta = 1 + \varepsilon e^{i\omega t} e^{-\sqrt{i\omega P_r} y} + \varepsilon^2 e^{2i\omega t} e^{-\sqrt{2i\omega P_r} y}$$

and either from Equations (12)-(14) under the respective boundary conditions (21), and substituting the solutions into Equation (9). We have the velocity distribution.

$$u = \frac{S_c K + G_0}{G S_c K} - \frac{y^2}{2} - \frac{G_0 e^{-\sqrt{S_c K} y}}{G S_c K}$$

$$+ \varepsilon e^{i\omega t} \left(a_1 e^{-\sqrt{i\omega} y} - \frac{e^{-\sqrt{i\omega P_r} y}}{i\omega (P_r - 1)} - \frac{G_0 e^{-\sqrt{S_c (K + i\omega)} y}}{G \left[S_c (K + i\omega) - i\omega \right]} \right)$$

$$+ \varepsilon^2 e^{2i\omega t} \left(a_1 e^{-\sqrt{2i\omega} y} - \frac{e^{-\sqrt{2i\omega P_r} y}}{2i\omega (P_r - 1)} \right.$$

$$\left. - \frac{G_0 e^{-\sqrt{S_c (K + 2i\omega)} y}}{G \left[S_c (K + 2i\omega) - 2i\omega \right]} \right)$$

by knowing velocity, temperature, and concentration profiles, it is interesting to study about local and average values of skin friction. In non-dimensional quantities, the skin friction

$$\tau = \frac{G_0 \sqrt{S_c K}}{G S_c K} + \varepsilon e^{i\omega t} \left(-a_1 \sqrt{i\omega} + \frac{\sqrt{i\omega P_r}}{i\omega (P_r - 1)} \right.$$

$$\left. + \frac{G_0 \sqrt{S_c (K + i\omega)}}{G \left[S_c (K + i\omega) - i\omega \right]} \right)$$

$$+ \varepsilon^2 e^{2i\omega t} \left(-a_2 \sqrt{2i\omega} + \frac{\sqrt{2i\omega P_r}}{2i\omega (P_r - 1)} \right.$$

$$\left. + \frac{G_0 \sqrt{S_c (K + 2i\omega)}}{G \left[S_c (K + 2i\omega) - 2i\omega \right]} \right)$$

where,

$$a_1 = \frac{1}{i\omega(P_r - 1)} + \frac{G_0}{G\left(S_c\left(K + i\omega\right) - i\omega\right)},$$

$$a_2 = \frac{1}{2i\omega(P_r - 1)} + \frac{G_0}{G\left(S_c\left(K + 2i\omega\right) - 2i\omega\right)}$$

RESULTS AND DISCUSSION

In order to get a physical understanding of the problem and for purpose of discussing the results, numerical calculations have been performed for the concentrationvelocity, temperature, rate of mass transfer, skin friction and rate of heat transfer. The results are represented graphically in Figures 1-10. The Prandtl number, $P_r = 0.71$ corresponds to air. The Grashof number, $G > 0$ corresponds to cooling of the plate by free convection currents, and Grashof number $G < 0$ corresponds to heating of the plate by free convection currents [3-6,12-19]. The mass diffusion Equation (8) can be adjusted to meet these circumstances if one takes, $k > 0$ for the destructive reaction, $k = 0$ for no reaction and $k < 0$ for the generative reaction. The effect of Prandtl number is very important in temperature profiles. There is a decrease in temperature due to increasing values of the Prandtl number.

The numerical values of the concentration profiles are computed and plotted in Figure 1 for different values of the chemical reaction parameter. Chemical reaction increases the rate of interfacial mass transfer. The reaction reduces the concentration. It is observed that the concentration increases during generative reaction and decreases in destructive reaction. For the case of destructive reaction increasing values of K leads to a fall in velocity profile as shown inFigure 2. A temporal maximum of velocity profiles is clearly seen for decreasing values of K. For generative reaction, a fall in velocity is observed for increasing K. This is due to the fact that as $K < 0$, the last term in the momentum equation becomes positive and plays a crucial role. Time required reaching steady state increases as K decreases. Here the difference between temporal velocity and steady state velocity is not clear as in the case of $K > 0$.

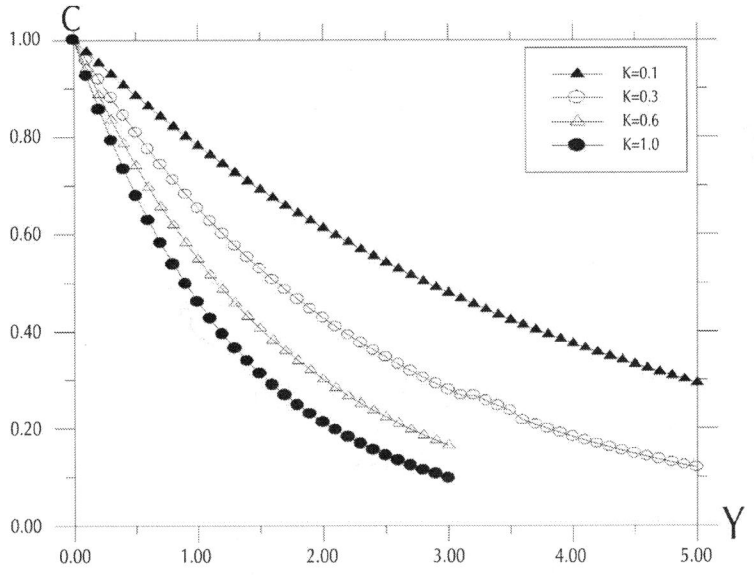

Figure 1: Concentration profile against spanwise coordinate Y for different values of K, with $P_r = 0.71$, $t = 0.2$, $= 5.0$, $G = 5.0$, $G_0 = 5.0$, $S_c = 0.6$.

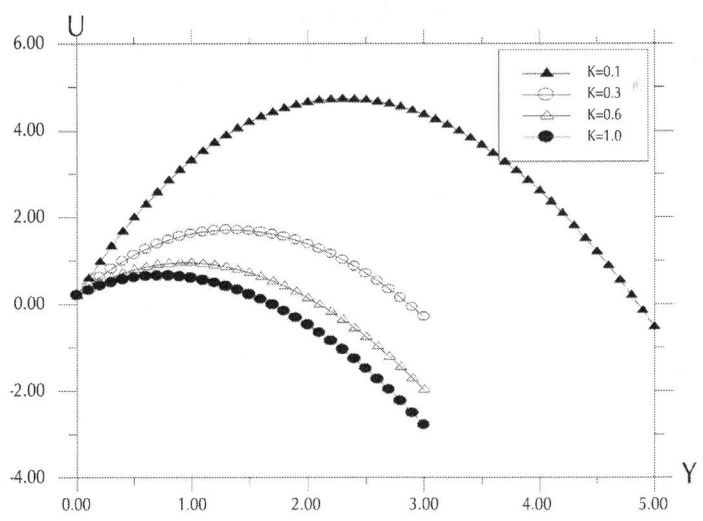

Figure 2: Velocity profile against spanwise coordinate Y for different values of K, with $P_r = 0.71$, $t = 0.2$, $= 5.0$, $G = 5.0$, $G_0 = 5.0$, $S_c = 0.6$.

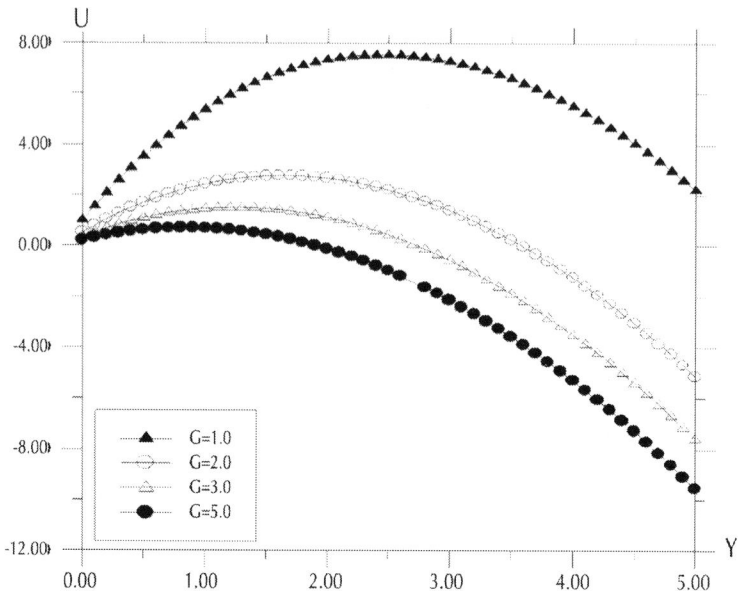

Figure 3: Velocity profile against spanwise coordinate Y for different values of Grashof number (G) with $P_r = 0.71$, t = 0.2, = 5.0, k = 0.2, $G_0 = 2.0$, $S_c = 0.6$.

The velocity profiles for different values of Grashof number G are described in Figure 3. It is observed that an increase in G leads to a decrease in the values of velocity. In addition, the curves show that the peak value of velocity increases rapidly near the wall of the plate as Grashof number decreases, and then decays to the relevant free stream velocity. The velocity profiles for different values of modified Grashof number G_0 are described in Figure 4. It is observed that an increase in G_0 leads to a decrease in the values of velocity. In addition, the curves show that the peak value of velocity increases rapidly near the wall of the plate as modified Grashof number decreases, and then decays to the relevant free stream velocity.

The transient concentration profiles for different Schmidt number is shown in Figure 5. It is observed that the concentration increases with decreasing Schmidt number. As the Schmidt number increases, the mass transfer rate increases and hence the concentration profiles decreases. The velocity profiles for different values of Schmidt number is shown in Figure 6. It is observed that the velocity decreases with increasing values of the Schmidt number.

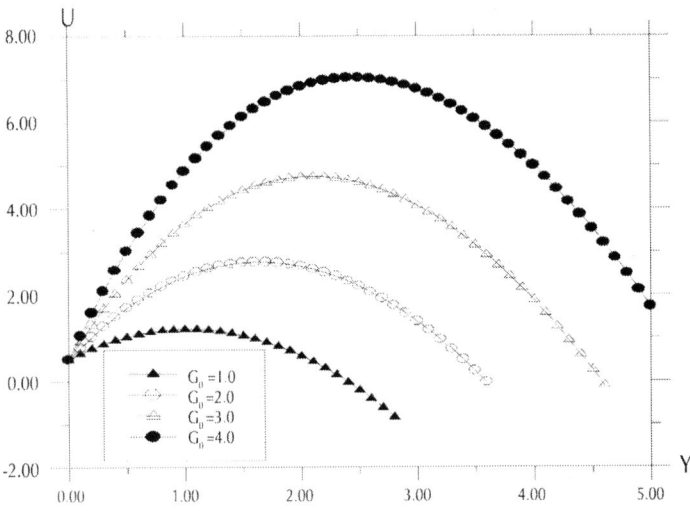

Figure 4: Velocity profile against spanwise coordinate Y for different values of modified Grashof number (G_0) with $P_r = 0.71$, t = 0.2, = 5.0, k = 0.2, G = 2.0, S_c = 0.6.

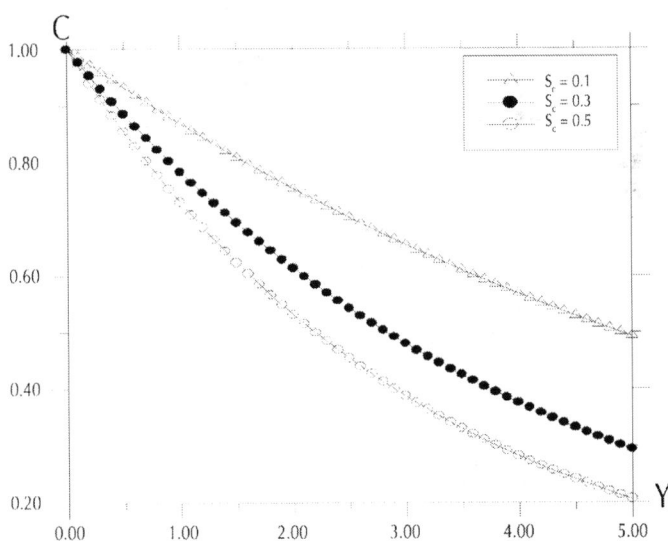

Figure 5: Concentration profile against spanwise coordinate Y for different values of Schmidt number (S_c), with $P_r = 0.71$, t = 0.2, = 5.0, G = 2.0, G_0 = 2.0.

The effects of buoyancy ratio parameter for both aiding $(G/G_0 > 0)$ as well as opposing $(G/G_0 < 0)$ are shown in Figure 7. It is observed that the velocity increases in the presence of opposing flows and decreases with aiding flows. Local skin friction values are plotted in Figure 8 against the Grashof number G. They are increasing for decreasing values of S_c. Increasing values of S_c and P_r give rise to lower shear stress. Since increasing S_c and P_r gives thicker velocity profiles which in turn give lower skin friction values. As the buoyancy ratio parameter increases, higher skin friction is observed. For generative reaction, shear stress decreases as reaction parameter decreases. A similar situation is noted for destructive reaction.

Local skin friction values are plotted in Figures 9, 10 against the Grashof number G. They are increasing for decreasing values of K. As the buoyancy ratio parameter increases, higher skin friction is observed. For generative reaction, shear stress decreases, as reaction parameter decreases. A similar situation is noted for destructive reaction.

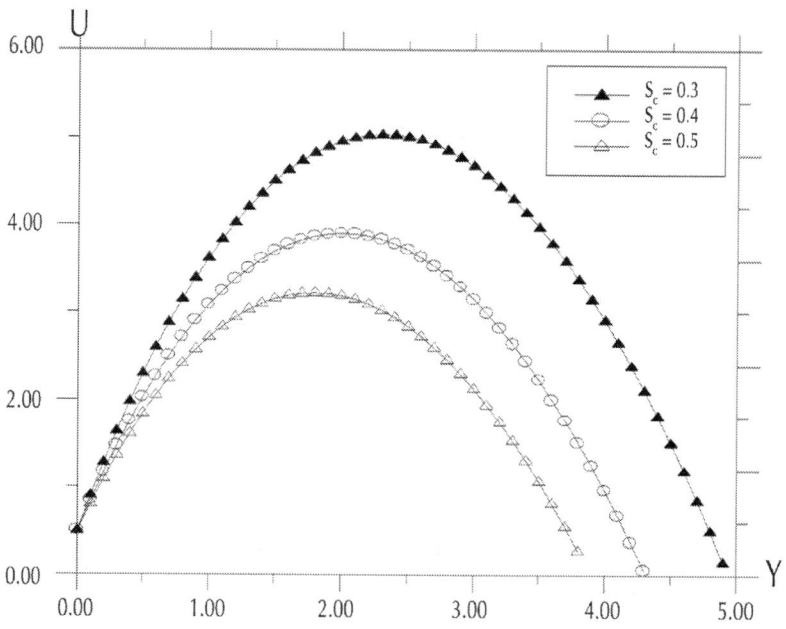

Figure 6: Velocity profile against spanwise coordinate Y for different values of Schmidt number (S_c) with $P_r = 0.71$, $t = 0.2$, $= 5.0$, $k = 0.2$, $G = 2.0$, $G_0 = 2.0$.

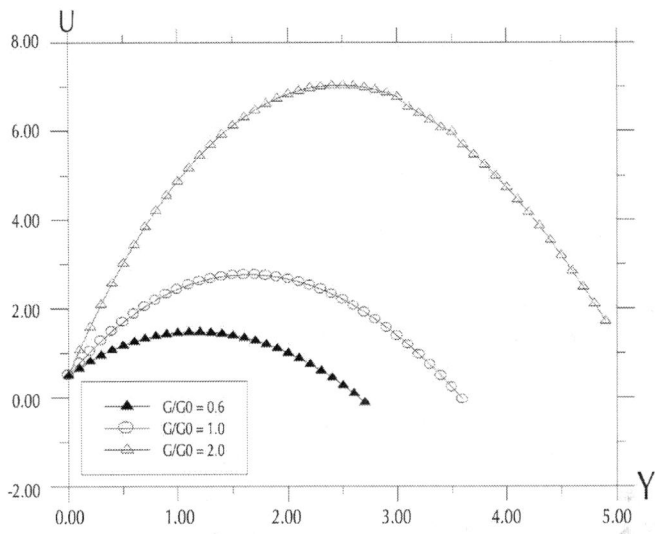

Figure 7: Velocity profile against spanwise coordinate Y for different values of (G/G_0), with $P_r = 0.71$, $t = 0.2$, $= 5.0$, $K = 0.2$, $S_c = 0.6$.

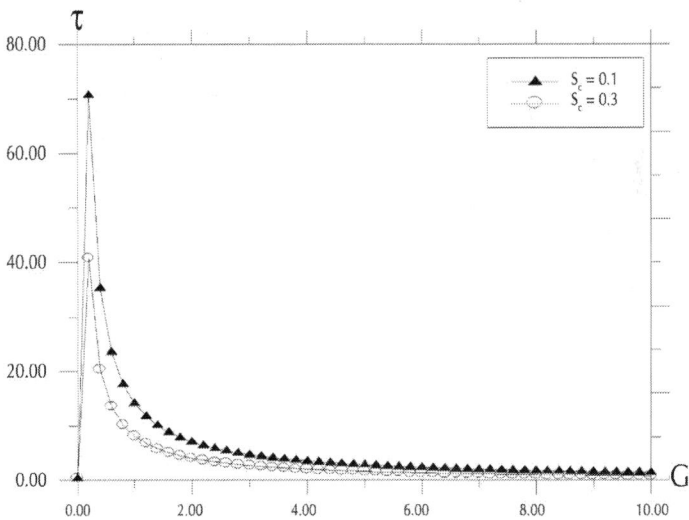

Figure 8: Skin Friction profile against Grashof number (G) for different values of Schmidt number (S_c) with $P_r = 0.71$, $t = 0.2$, $= 5.0$, $K = 0.2$, $y = 1$.

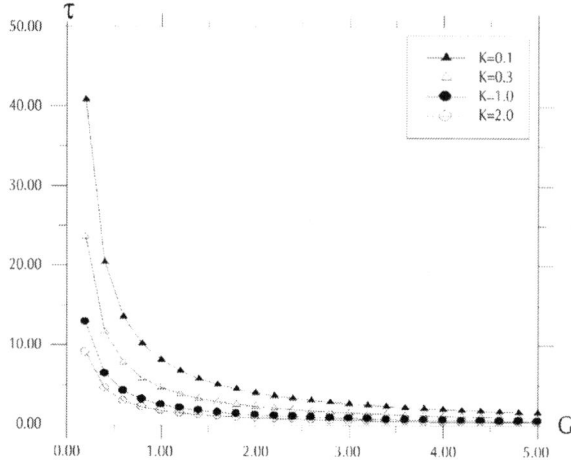

Figure 9: Skin friction profile against Grashof number (G) for different values of K, with $P_r = 0.71$, t = 0.2,   = 5.0, $S_c = 0.6$, y = 1.0.

CONCLUSIONS

A detailed numerical study has been carried out for the flow past an impulsively started infinite vertical plate with variable temperature and mass diffusion. These circumstances are of interest in several manufacturing processes.

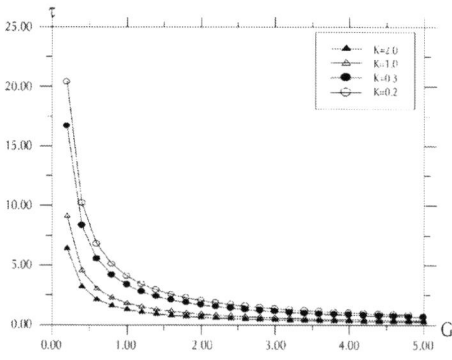

Figure 10: Skin Friction profile against modified Grashof number for different values of K, with $P_r = 0.71$, t = 0.2,   = 5.0, $S_c = 0.6$, y = 1.0.

The dimensionless governing equations are solved by a perturbation technique. Numerical evaluations of the closed form results were performed and some graphical results were obtained to illustrate the details of the flow and heat and mass transfer characteristics and their dependence on some of physical parameters. The study has been compared with available exact solution the literature and they are found to be in good agreement. It is observed that, the concentration increases during generative reaction and decreases in destructive reaction. The concentration increases with decreasing Schmidt number. The effect of increasing values of K leads to a fall in velocity profiles. The velocity decreases with increasing values of the Schmidt number. An increase in modified Grashof number leads to an increase in velocity profiles. The skin friction increases with decreasing Schmidt number. In generative reaction the skin friction decreases and in destructive reaction the skin friction increases.

ACKNOWLEDGEMENTS

Appreciation is extended to the referees for their constructive and helpful comments. These led to improvements in the revised paper.

REFERENCES

1. F. A. Bottemanne, "Theoretical Solution of Simultaneous Heat and Mass Transfer by Free Convection about a Vertical Flat Plate," Applied Scientific Research, Vol. 25, No. 1, 1972, pp. 137-149. doi:10.1007/BF00382290

2. T. S. Chen and C. F. Yuh, "Combined Heat and Mass Transfer in Natural Convection along a Vertical Cylinder," International Journal of Heat and Mass Transfer, Vol. 23, No. 4, 1980, pp. 451-461. doi:10.1016/0017-9310(80)90094-0

3. B. C. Sakiadis, "Boundary-Layer Behavior on Continuous Solid Surfaces: I. Boundary-Layer Equations for TwoDimensional and Axisymmetric Flow," AIChE Journal, Vol. 7, No. 1, 1961, pp. 26-28. doi:10.1002/aic.690070108

4. B. C. Sakiadis, "Boundary-Layer Behavior on Continuous Solid Surfaces: II. The Boundary Layer on a Continuous Flat Surface,"

AIChE Journal, Vol. 7, No. 1, 1961, pp. 221-225.doi:10.1002/aic.690070211

5. V. M. Soundalgekar, "Effects of Mass Transfer and FreeConvection Currents on the Flow past an Impulsively Started Vertical Plate," Journal of Applied Mechanics, Vol. 46, No. 4, 1979, pp. 757-760. doi:10.1115/1.3424649

6. V. M. Soundalgekar, N. S. Birajdar and V. K. Darwhekar, "Mass-Transfer Effects on the Flow past an Impulsively Started Infinite Vertical Plate with Variable Temperature or Constant Heat Flux," Astrophysics and Space Science, Vol. 100, No. 1-2, 1984, pp. 159-164. doi:10.1007/BF00651593

7. P. L. Chambre and J. D. Young, "On the Diffusion of a Chemically Reactive Species in a Laminar Boundary Layer Flow," Physics of Fluids, Vol. 1, 1958, pp. 48-54.doi:10.1063/1.1724336

8. K. V. Ramanamurthy and V. M. Govinda Rao, Proceedings of the First National Heat and Mass Transfer Conference, Chennai, 1971.

9. U. N. Das, R. Deka and V. M. Soundalgekar, "Effects of Mass Transfer on Flow past an Impulsively Started Infinite Vertical Plate with Constant Heat Flux and Chemical Reaction," Forschung im Ingenieurwesen, Vol. 60, No. 10, 1994, pp. 284-287.doi:10.1007/BF02601318

10. P. Ganesan and H. P. Rani, "Transient Natural Convection along Vertical Cylinder with Heat and Mass Transfer," Heat and Mass Transfer, Vol. 33, No. 5-6, 1998, pp. 449-455.doi:10.1007/s002310050214

11. P. Ganesan and H. P. Rani, "On Diffusion of Chemically Reactive Species in Convective Flow along a Vertical Cylinder," Chemical Engineering and Processing, Vol. 39, No. 2, 2000, pp. 93-105. doi:10.1016/S0255-2701(99)00018-5

12. M. M. Abdelkhalek, "The Skin Friction in the MHD Mixed Convection Stagnation Point with Mass Transfer," International Communications in Heat and Mass Transfer, Vol. 33, No. 2, 2006, pp. 248-257. doi:10.1016/j.icheatmasstransfer.2005.09.008

13. M. M. Abdelkhalek, "Mixed Convection in a Square Cavity by a Perturbation Technique," Computational Materials Science, Vol. 42, No. 2, 2008, pp. 212-219.doi:10.1016/j.commatsci.2007.07.004

14. M. M. Abdelkhalek, "Hydromagnetic Stagnation Point Flow by a Perturbation Technique," Computational Materials Science, Vol. 42, No. 3, 2008, pp. 497-503.doi:10.1016/j. commatsci.2007.08.013

15. M. M. Abdelkhalek, "Heat and Mass Transfer in MHD Flow by Perturbation Technique," Computational Materials Science, Vol. 43, No. 2, 2008, pp. 384-391.doi:10.1016/j. commatsci.2007.12.003

16. M. M. Abdelkhalek, "Unsteady MHD Convection and Mass Transfer Flow of Micropolar Fluids past a Vertical Permeable Moving Plate with heat Absorption," Indian Journal of Physics, Vol. 80, No. 6, 2006, pp. 625-635.

17. M. M. Abdelkhalek, "Thermal Radiation Effects on Hydromagnetic Flow," Computer Assisted Mechanics and Engineering Sciences, Vol. 14, No. 3, 2007, pp. 471-484.

18. M. M. Abdelkhalek, "Radiation and Dissipation Effect on Unsteady MHD Micropolar Flow past an Infinite Vertical Plate in a Porous Medium with Time Dependent Suction," Indian Journal of Physics, Vol. 82, No. 4, 2008, pp. 415-434.

19. M. M. Abdelkhalek, "Heat and Mass Transfer in MHD Free Convection from a Moving Permeable Vertical Surface by a Perturbation Technique," Communications in Nonlinear Science and Numerical Simulation, Vol. 14, No. 5, 2009, pp. 2091-2102. doi:10.1016/j.cnsns.2008.06.001

20. S. A. Zarea, F. M. El-Fayez and M. M. A. Khalek, "Perturbation Technique Algorithm for Mixed Convection Flow in a Confined Saturated Porous Medium with Temperature," Arab Journal of Nuclear Sciences and Applications, Accepted, 2010.

21. B. Gebhart, Y. Jaluria, R. L. Mahajan and B. Sammakia, "Buoyancy-Induced Flows and Transport," Hemisphere Publishing Corporation, New York, 1988.

22. R. Muthucumaraswany and P. Ganesan, "Diffusion and First-Order Chemical Reaction on Impulsively Started infinite Vertical Plate with Variable Temperature," International Journal of Thermal Science, Vol. 41, No. 5, 2002, pp. 475-479.

Chapter 7

Microwave Plasma Enhanced Chemical Vapor Deposition of Carbon Nanotubes

Ivaylo Hinkov[1], Samir Farhat[2], Cristian P. Lungu[3],
Alix Gicquel[2], François Silva[2], Amine Mesbahi[2],
Ovidiu Brinza[2], Cornel Porosnicu[3], and Alexandru
Anghel[3]

[1]University of Chemical Technology and Metallurgy, Sofia, Bulgaria

[2]Laboratoire des Sciences des Procédés ET des Matériaux, CNRS, LSPM-UPR 3407, Université Paris 13, Villetaneuse, France

[3]National Institute for Laser, Plasma and Radiation Physics, Bucharest, Romania

ABSTRACT

Multi-walled carbon nanotubes (MWCNTs) were grown by plasma-enhanced chemical vapor deposition (PECVD) in a bell jar reactor. A mixture of methane and hydrogen (CH_4/H_2) was decomposed over Ni catalyst previously deposited on Si-wafer by thermionic vacuum arc

(TVA) technology. The growth parameters were optimized to obtain dense arrays of nanotubes and were found to be: hydrogen flow rate of 90 sccm; methane flow rate of 10 sccm; oxygen flow rate of 1 sccm; substrate temperature of 1123 K; total pressure of 10 mbar and microwave power of 342 Watt. Results are summarized and significant main factors and their interactions were identified. In addition a computational study of nanotubes growth rate was conducted using a gas phase reaction mechanism and surface nanotube formation model. Simulations were performed to determine the gas phase fields for temperature and species concentration as well as the surface-species coverage and carbon nanotubes growth rate. A kinetic mechanism which consists of 13 gas species, 43 gas reactions and 17 surface reactions has been used in the commercial computational fluid dynamics (CFD) software ANSYS Fluent. A comparison of simulated and experimental growth rate is presented in this paper. Simulation results agreed favorably with experimental data.

INTRODUCTION

Since their discovery by Iijima in 1991 [1] , carbon nanotubes have generated much interest due to their quasi one-dimensional structure and their unique combinations of electronic, field emission, mechanical and chemical properties coupled with the new ability to grow them aligned on a substrate. This opened unlimited possibilities of applications such as field emitters, sensors, high-density energy storage devices, photonic crystals, active media for lasers, non-linear optical media etc... To grow vertically-aligned nanotubes, chemical vapor deposition (CVD) has emerged as a key technique. Indeed, contrarily to the arc, laser and HiPCO processes where the nanotubes are produced separately, purified and then manipulated for producing devices [2], CVD allows spatially controlled and highly functional components in (2D) and (3D) architecture opening the way to produce selfassembly devices with higher packing density and performances [3]. In addition, CVD offers low-temperature and large-scale production possibilities. In CVD systems, a thin catalyst layer is first deposited on silicon wafer by a separate physical vapor deposition PVD technique. When heated, the continuous catalyst layer disaggregates and forms small particles, with size controlled by the layer thickness in the range of 1 to 10 nm [4]. Then, the growth of nanotubes occurs through catalytic decomposition of a carbon gas source over the catalyst. The nanotube characteristics such as diameter, density, Single-walled SWNT versus Multi-walled MWNT depend on the size of these particles but also on the gas feedstock activation technique. Two distinct activation routes emerged, 1) thermally via an oven or hot-filament heating and 2) plasma enhanced chemical vapor deposition (PECVD) via DC, RF or microwave discharges. Plasma activation has the advantage to prevent thermal damage to the substrate allowing lower operating temperatures and better nanotubes vertical positioning due to the presence of an electric field normal to the substrate [5].

In the present work, we used thermionic vacuum arc (TVA) technology to produce uniform nickel layers of ~1 nm thickness. Then, a mixture of methane, hydrogen and oxygen ($CH_4/H_2/O_2$) was used to produce carbon nanotubes. Indeed, the addition of a controlled amount of a weak oxidizer as oxygen or water into the growth ambient of CVD was reported to significantly enhance the activity and lifetime of the catalyst resulting in efficient nanotubes growth [6] [7] .

EXPERIMENTAL

Nickel films deposited on silicon substrates were prepared using thermionic vacuum arc (TVA) technology developed at NILPRP Bucharest [8] -[11] . The coating device consists of a tungsten filament surrounded by an electron focusing Wehnelt cylinder heated by an external high current source as cathode and an anode made of nickel. For ignition and maintaining the TVA arc two circuits are necessary: 1) for the heating of the cathode filament, where a relatively low voltage source (0 - 24 V) provides a 10 - 150 A current and 2) for the running up of the arc discharge, being used for this an adjustable source of high voltage (0 - 4 kV) and a current up to 3 A. The electrons coming from the cathode heats up and evaporates the anode and pure Ni plasma is ignited by applying a high dc voltage on the anode as illustrated in Figure 1(a). The deposition chamber of Figure 1(b) has been under a residual pressure of 3×10^{-6} torr before the beginning of the coatings. For plasma ignition, the TVA gun filament has been heated with a 60 A current and at an alternative voltage of 20 V. The continuous voltage has been applied on the anode with an increasing rate of approximately

1000 V/min, being followed by the focusing process of the electron beam by the Wehnelt cylinder on the anode crucible. When the powder in the crucible has melt, the applied voltage was adjusted in order to ensure the ignition of the discharge in the Ni vapors. A stable discharge shown in Figure 1(c), is obtained and the film thickness measured during all the duration of the deposition process with a quartz balance equipment. The deposition has been interrupted when the thickness of 1 nm is reached. At this step, the anode voltage and the applied current to the TVA gun filament have been reduced to zero and the sample kept in the deposition chamber, under high vacuum for about 120 minutes to slowly cool down. Finally, Ni/Si substrates with ~1 nm nickel thickness were obtained in 18 - 20 s, due to a fine control of the deposition rate of ~0.05 nm/s.

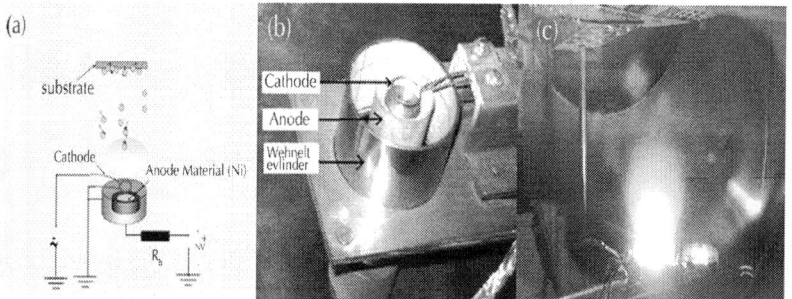

Figure 1: Thermionic vacuum arc TVA set-up for catalyst deposition. (a) Principle, (b) deposition chamber and (c) plasma running in Ni vapors.

For nanotube growth, we used a 10 cm diameter silica bell jar low pressure reactor activated by a microwave electric field (Figure 2) and developed originally to CVD diamond growth [12] [13] . The input gases ($CH_4/H_2/O_2$) with mass flow rates controlled electronically were injected in the reactor and exit via the reactor pumping system. The Ni/Si substrate is held in a resistance boat made in molybdenum and electrically heated to a temperature ranging from 973 to 1123 K. During all the experiment, substrate temperature was monitored by an optical pyrometer. The reactor utilizes 1.2 kW SAIREM microwave generator operating at 2.45 Ghz. The electromagnetic waves are generated, guided in a rectangular wave guide and applied inside the cavity delimited by Faraday cage (Figure 2(a)). The short-circuit piston at the end of the wave guide helped to create stationary waves and to situate the maximum of the electric field near the substrate. Input power was varied with the pressure simultaneously in order to hold plasma volume constant (Figure 2(b)). Efficient operation is assumed with good microwave coupling, and minimal radial diffusion to the quartz enclosure thereby leading to greater discharge stability and better plasma uniformity. As shown in Figure 2(b), a quasi-hemispherical active plasma zone of radius of 2.5 cm is created near the substrate. The function of this zone is to produce the charged and radical species that diffuse to the catalyst particle and contribute to nanotube growth. For microwave PECVD nanotube synthesis we developed an experimental protocol composed by three steps: 1) thermal annealing of Ni/Si substrates, 2) hydrogenation of Ni catalyst, 3) nanotube growth.

For this protocol, the essential parameters were optimized using Taguchi design method of Table 1, with 3 factors and 2 levels. These parameters are namely, substrate temperature, hydrogen flow rate, and total pressure. This last parameter was coupled with the total input microwave power to hold the plasma volume constant. For all the 8 experiments, the silicon substrate covered by 1 nm thick nickel was first annealed in vacuum at specified temperature, then nickel catalyst particles was reduced using hydrogen plasma for 10 minutes. Finally, 10 sccm of methane were introduced to grow nanotubes during 20 minutes.

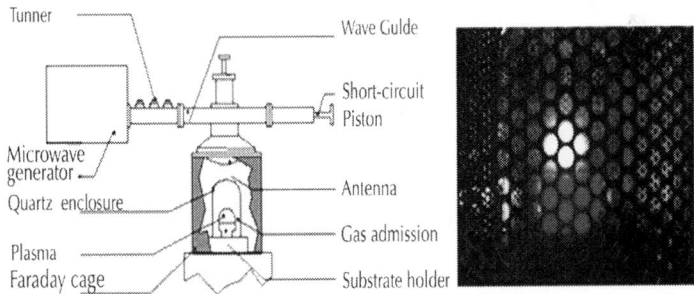

Figure 2: PECVD Bell jar reactor. (a) Scheme and (b) plasma picture through the Faraday cage during nanotubes growth.

Table 1: Factors and levels for the PECVD synthesis of nanotubes. Pressure and microwave power are coupled

Experiment No.	Substrate Temperature Tsub (K)	H2 Flow rate Q_{H_2} (sccm)	CH4 Flow rate Q_{CH_4} (sccm)	Total pressure P (mbar)	Microwave power Pmw(Watt)
E1	973	10	10	10	342.82
E2	1123	10	10	10	342.82
E3	973	10	10	40	814.54
E4	1123	10	10	40	814.54
E5	973	100	10	10	342.82
E6	1123	100	10	10	342.82
E7	973	100	10	40	814.54
E8	1123	100	10	40	814.54

After removing samples from the reactor, they were analyzed by scanning electron microscopy SEM LEO 440. In addition, surface morphology of the substrates, before and after thermal annealing was examined by atomic force microscopy AFM D3100, Nanoscope NS3.

MODELING APPROACH

A two-dimensional simulation of the PECVD process was performed in order to compare the theoretical prediction and the experimental measurements. Computational fluid dynamics (CFD) modeling evaluations were made for the temperature and species concentrations profiles as well as for the carbon nanotube growth rate in the reactor. Simulations were performed by using the commercial software ANSYS Fluent version 12.

Geometry and Assumptions

The 2D computational domain shown in Figure 3 includes the PECVD reactor quartz enclosure, containing the substrate holder and the plasma zone. From experimental observations, plasmas are most intense along with the top edge of the substrate holder. The microwave plasma volume used in the present simulations was estimated from visual observations under experimental growth conditions in the reactor. The geometry was created using ANSYS Design Modeler and the mesh is generated using ANSYS Meshing application. The chosen dimensions of the reactor refer to the experimental setup. The grid is composed of an unstructured triangular mesh. Because of the strong temperature and concentrations gradients near the substrate where the CNT are grown, a condensed quadrilateral mesh refinement was applied on this region. After numerous checks for grid sensitivity and mesh constraints, the total number of elements is 4583, the final grid has 2487 nodes with a grid skewness maximum of 0.78. To understand the effect of the macroscopic process parameters on carbon nanotubes growth rate, a 2D model has been developed based on the following assumptions:

- The plasma is in Local Thermodynamic Equilibrium (LTE). This assumption allows us to define a unique temperature of all plasma species in localized areas in the plasma;
- Laminar flow: it is characterized by relatively low values of the Reynolds number caused by small inlet flow rate;
- The plasma is modeled using a steady state time formulation;
- Axisymmetric physical domain;
- The radiative losses are neglected;
- Only neutral species are involved in the gas-phase and surface chemistry.

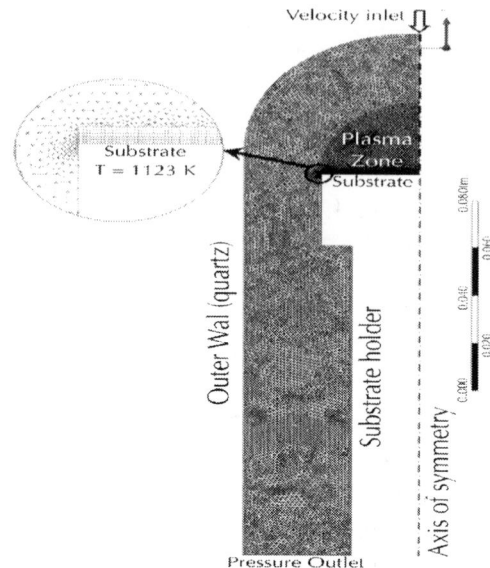

Figure 3: Two-dimensional computational domain.

These assumptions lead to the following limitations: The existence of only neutral species affects the accuracy of the calculated concentrations of considered species. We expect the influence of the charged species on carbon nanotubes growth rate to be minor because their molar fraction is not great. The major species formed in the plasma are H, H_2, CH_4, CH_3, C_2H_2, and C_2H_4 and the gas phase chemistry in the plasma is dominated by the neutral species [14]. Furthermore, assuming LTE reduces the complexity of the mixture considerably.

Gas-Phase Chemistry

In the methane/hydrogen plasma, different species are created due to chemical reactions. The considered gasphase chemistry involves 13 neutral species and consists of 43 reactions leading to the conversion of CH_4. This reaction set is given in Table 2. The gas phase chemistry model describes homogeneous reactions that influence the species concentration distribution near the deposition surface through the production/destruction of chemical species in the gas phase. Each reaction is assumed to be reversible. The temperature dependence of the forward rate constants is usually described through a modified Arrhenius type of expression.

Surface Chemistry

The surface chemistry model used in the present study describes the reactions and other processes that take place at the substrate surface, involving both gaseous species impinging on the surface, adsorbed molecules, atoms and free sites. These surface processes lead to the growth of solid carbon nanotubes. Actually, surface reactions in the PECVD process are not fully understood. The proposed surface reaction mechanism consists of 18 heterogeneous reactions involving vacant surface sites S_{Ni} on a nickel catalyst particle, 7 surface species ($CH_4(s)$, $CH_3(s)$, $C_1H_2(s)$, $CH(s)$, $C(s)$, $H(s)$, CNT) and 5 gaseous species (C_2H_2, CH_4, CH_3, H_2, H). These reactions include surface site adsorption/desorption, hydrogen abstraction/addition, and carbon diffusion toward a carbon nanotube growth edge (Table 3).

The rate of deposition is governed by both chemical kinetics and the diffusion rate from gas to the surface. The reactions create sources of chemical species in the bulk phase and determine the rate of deposition of surface species.

Following the above considerations, the growth of MWNT on the substrate is expected to occur as follows. The plasma generates vapors and provides carbon contamination to the nickel particles. These particles having suitable temperature and size will be the sites of CNT growth.

The surface structure is associated with a surface site density (given in mol·cm^{-2}) required to evaluate the surface growth rate of MWNT. Since experimental determination of site density is difficult, we used the value for the reconstructed diamond (100) surface = 2.61×10^{-9} as an upper limit. In the specific case of nanotube growth in PECVD reactor, the surface site density is certainly much lower than this value. It could be estimated from the substrate density of catalytic nickel particles and the concentration of the surface sites occupied by C atoms on each nanoparticle.

Initial and Boundary Conditions

The temperature at the outer boundary walls and the substrate holder except for the substrate surface was fixed to 400 K. The temperature of the substrate surface was much higher, at 1123 K.

The inlet conditions of the simulations were derived from the experimental conditions i.e., volume %: 10% CH_4 and 90% H_2. The gas mixture was initialized to a uniform temperature of 298 K. The gas velocity was specified as a uniform inflow condition with vertical upward velocity of 0.1326 m/s at a temperature of 298 K. The initial inlet mole fractions for all species were calculated using a thermochemical model based on Chemkin software in 0D [31]. The gas outlet was specified as a pressure outlet. The initial pressure inside the reactor was fixed at 10 mbar.

Computational Procedure

2D reactor simulation including coupled momentum, heat and species transfer was performed by using the CFD code ANSYS Fluent. It utilizes the finite volume method to solve the governing equations, i.e., conservation of total mass, momentum, and energy, and the individual species conservation equations. The reactive flow is modeled using the 2D axisymmetric laminar finite-rate model, including the above-mentioned volumetric and surface reactions. The Simple method for pressure-velocity coupling and the second order upwind scheme to interpolate the variables on the surface of the control volume were selected.

Table 2: Gas-phase reactions

No.	Gas Phase Reactions	A*	β (-)	Ea(cal/mol)	Ref.
1.	H + H + M = H2 + M	1.00 × 1018	−1.0	0.0	[15]
	H2 Enhanced by 2.0				
2.	H + H + H2 = H2 + H2	9.20 × 1016	−0.6	0.0	[15]
3.	CH3 + CH3 (+M) = C2H6 (+M)	9.22 × 1016	−1.174	636.0	[16]
	Low pressure limit: 1.14 × 1036 −5.246 1705.0				
	TROE centering: 0.405 1120.0 69.6 1.0 × 1015				

	H2 Enhanced by 2.0				
4.	CH3 + H (+M) = CH4 (+M)	2.14×10^{15}	−0.4	0.0	[17]
	Low pressure limit: 3.31×10^{30} −4.0 2108.0				
	TROE centering: 0.0 1.00×10^{-15} 1.0×10^{-15} 40				
	H2 Enhanced by 2.0				
5.	CH4 + H = CH3 + H2	2.20×10^{4}	3.0	8750.0	[18]
6.	CH3 + H = CH2 + H2	9.00×10^{13}	0.0	15100.0	[18]
7.	CH3 + M = CH + H2 + M	6.90×10^{14}	0.0	82469.0	[19]
8.	CH3 + M = CH2 + H + M	1.90×10^{16}	0.0	91411.0	[19]
9.	CH2 + H = CH + H2	1.00×10^{18}	−1.56	0.0	[18]
10.	CH2 + CH3 = C2H4 + H	4.00×10^{13}	0.0	0.0	[18]
11.	CH2 + CH2 = C2H2 + H + H	4.00×10^{13}	0.0	0.0	[18]
12.	CH2(S) + M = CH2 + M	1.00×10^{13}	0.0	0.0	[18]
	H2 Enhanced by 12.0				
	C2H2 Enhanced by 4.0				
13.	CH2(S) + CH4 = CH3 + CH3	4.00×10^{13}	0.0	0.0	[18]
14.	CH2(S) + C2H6 = CH3 + C2H5	1.20×10^{14}	0.0	0.0	[18]
15.	CH2(S) + H2 = CH3 + H	7.00×10^{13}	0.0	0.0	[18]
16.	CH2(S) + H = CH + H2	3.00×10^{13}	0.0	0.0	[18]

17.	$CH_2(S) + CH_3 = C_2H_4 + H$	2.00 × 1013	0.0	0.0	[18]
18.	$CH + H = C + H_2$	1.50 × 1014	0.0	0.0	[18]
18.	$CH + CH_2 = C_2H_2 + H$	4.00 × 1013	0.0	0.0	[18]
20.	$CH + CH_3 = C_2H_3 + H$	3.00 × 1013	0.0	0.0	[18]
21.	$CH + CH_4 = C_2H_4 + H$	6.00 × 1013	0.0	0.0	[18]
22.	$C + CH_3 = C_2H_2 + H$	5.00 × 1013	0.0	0.0	[18]
23.	$C + CH_2 = C_2H + H$	5.00 × 1013	0.0	0.0	[18]
24.	$C_2H_6 + CH_3 = C_2H_5 + CH_4$	5.50 × 10−1	4.0	8300.0	[18]
25.	$C_2H_6 + H = C_2H_5 + H_2$	5.40 × 102	3.5	5210.0	[18]
26.	$C_2H_5 + H = C_2H_4 + H_2$	1.25 × 1014	0.0	8000.0	[15]
27.	$C_2H_5 + H = CH_3 + CH_3$	3.00 × 1013	0.0	0.0	[20]
28.	$C_2H_5 + H = C_2H_6$	1.00 × 1014	0.0	0.0	[15]
29.	$C_2H_4 + H = C_2H_3 + H_2$	3.36 × 10−7	6.0	1692.0	[21]
30.	$C_2H_4 + CH_3 = C_2H_3 + CH_4$	6.62	3.7	9500.0	[15]
31.	$C_2H_4 + H (+M) = C_2H_5 (+M)$	1.08 × 1012	0.454	1822.0	[22]
	Low pressure limit: 1.112 × 1034 −5.0 4448.0				
	TROE centering: 1.0 1.00 × 10−15 95.0 200.0				
	H2 Enhanced by 2.0				

*Units for A depend on the reaction order but are defined in terms of mol, cm^3 and s.

Table 3: Surface reactions

No.	Surface Reactions	A*	B (-)	Ea(cal/mol)	Ref.
S1.	H2 + SNi + SNi → H(s) + H(s)	0.01**	0.0	0.0	[28]
S2.	H(s) + H(s) → H2 + SNi + SNi	2.545 × 1019	0.0	19379.0	[28]
S3.	CH4 + SNi → CH4(s)	0.008**	0.0	0.0	[28]
S4.	CH4(s) → CH4 + SNi	8.705 × 1015	0.0	8962.0	[28]
S5.	CH4(s) + SNi → CH3(s) + H(s)	3.700 × 1021	0.0	13770.8	[28]
S6.	CH3(s) + H(s) → CH4(s) + SNi	6.034 × 1021	0.0	14701.6	[28]
S7.	SNi + CH3 → CH3(s)	5.000 × 1012	0.0	0.0	[29]
S8.	CH3(s) + H = C1H2(s) + H2	2.800 × 107	2.0	7700.0	[29]
S9.	C1H2(s) + H = CH(s) + H2	2.800 × 107	2.0	7700.0	[29]
S10.	CH(s) + H = C(s) + H2	2.800 × 107	2.0	7700.0	[29]
S11.	C1H2(S) + H ⊠ SNi + CH3	3.000 × 1013	0.0	0.0	[29]
S12.	CH(s) + SNi ⊠ H(s) + C(s)	3.700 × 1021	0.0	4486.8	[28]
S13.	C(s) + H(s) ⊠ CH(s) + SNi	4.562 × 1022	0.0	38448.5	[28]
S14.	SNi + H ⊠ H(s)	1.000 × 1013	0.0	0.0	[29]
S15.	H(s) + H ⊠ SNi + H2	1.300 × 1014	0.0	7.3	[29]
S16.	2SNi + C2H2 ⊠ 2C(s) + H2	7.700 × 1010	0.0	67160.0	[30]
S17.	C(s) = CNT + SNi	1.300 × 1012	0.0	31104.0	[29]

S18.	C(s) + CNT = 2CNT + SNi	1.300 × 1012	0.0	31104.0	[29]

*Units for A depend on the reaction order but are defined in terms of mol, cm³ and s.
**Sticking coefficient.

The model requires knowledge of the thermo-chemical and transport properties of the gases in the reactor chamber. Thermo-chemical properties of the gas species as a function of temperature have been taken from CHEMKIN thermodynamic database [32] - [33]. Temperature and species dependence was imposed in calculations of thermodynamic and transport properties. The required Lennard-Jones parameters for many CVD gases can be found in e.g. Ref. [34]. Viscosity of the individual species was calculated by using kinetic theory.

The mixture viscosity was calculated using ideal gas mixing law. Thermal conductivity for individual species was calculated using kinetic theory. Specific heat capacity of individual species was calculated using piecewise-polynomial approximation.

To account for the plasma heating from the microwave power, a constant volumetric heat source, applied in the plasma zone, was included in the governing energy equation. It was calculated from the input plasma power and the estimated plasma volume. For input plasma power of 342.8 W, the calculated power density was 0.634 × 10^7 W/m³. The heat source using CH_4 and H_2 as a medium creates plasma field with a temperature over 2000 K.

The solution was initialized from the inlet. It was monitored approximately up to 20,000 iterations with residual convergence fixed between 1×10^{-3} and 1×10^{-5}.

The calculated rate of production s_{CR} expressed in moles/cm²/s is converted to linear nanotube growth rate G in m/s by using nanotube bulk mass density ρ_{CNT} = 2.20 g/cm³ and molecular weight M_{CNT} = 12.01 g/mol using the equation:

$$G = s_{CR} \cdot M_{CNT} / \rho_{CNT}$$

(1)

RESULTS AND DISCUSSION

Experimental Results

Atomic force microscopy (AFM) of the as produced by TVA Ni/Si substrates and annealed at 850°C during 20 minutes then hydrogenated with pure hydrogen plasma during 10 minutes was carried out to determine the surface morphology. Figure 4 shows AFM images of the substrate before (a and b) and after (c and d) annealing.

The change in the root mean square roughness RMS was from 1.864 nm before annealing to 3.485 nm after annealing. This higher rough surface clearly indicates an agglomeration of individual nickel clusters. After thermal annealing of Ni/Si substrates at the consigned temperature during 20 minutes, we followed the same experimental protocol for all the 8 experiments of Table 1.

First, substrates are treated with pure hydrogen plasma during 10 minutes then 10 sccm of methane was added into the mixture for 20 minutes. Scanning Electron Microscope indicates that no or few nanotubes were found for samples at 700°C E1, E3, E5 and E5 suggesting that 700°C is the lower growth temperature. Long spiral or helical nanotubes were observed for sample E2 and shorter nanotubes at different densities were found in samples E4, E6 and E8. Since carbon diffusion in bulk Ni is characterized by a large energy, namely, activation energy of 1.4 eV (33 kcal/mole) [35], we can estimate the diffusion coefficient of carbon in nickel using the following equation at lower and upper temperatures.

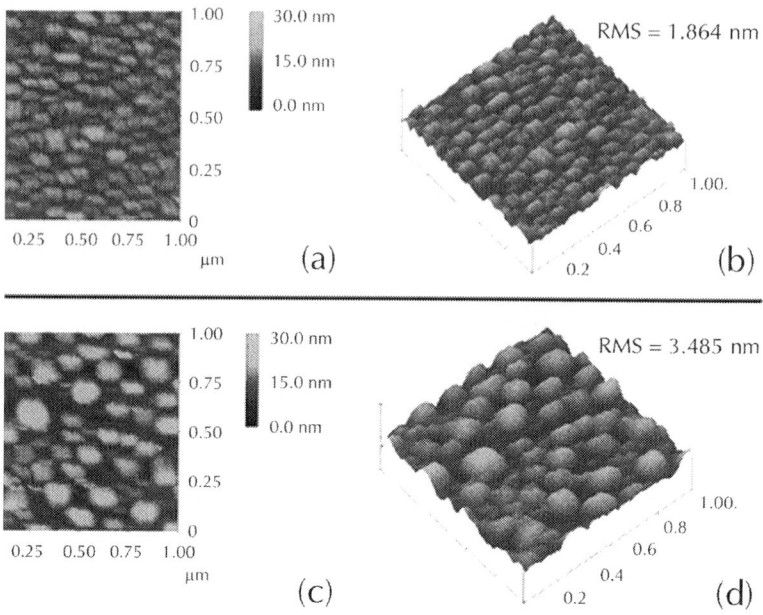

Figure 4: AFM images of the substrate before ((a) and (b)) and after ((c) and (d)) annealing.

$$D_{C-Ni} = 0.1 \times \exp\left(-33000/RT\right)$$

(2)

At 700°C, D_{C-Ni} ~3.9 × 10⁻⁹ m²·s⁻¹ and at 850°C, D_{C-Ni} ~3.8 × 10⁻⁸ m²·s⁻¹, hence carbon diffusion in the catalyst is increased by one order of magnitude.

To explain the role of the combined parameter on growth, we calculated the plasma kinetics using a thermochemical model based on Chemkin software in 0D [36]. This model actually includes 119 species in C/H system with atoms and molecules including Polycyclic aromatic hydrocarbons (PAHs), ions and electrons and 336 chemical reactions [13] [37]. Figure 5, shows the calculated hydrogen atom mole fraction in the plasma for the different conditions of Table 1. It clearly indicates that hydrogen atoms excess is not suitable for nanotube growth in microwave plasma systems.

In extreme case E3, SEM pictures revealed a strong etching of the substrate. Based on these observations, the best combination of parameters are those of experiment E2 which will be retained to explore the effect of oxygen. It was reported by several workers that addition of a controlled amount of oxygen or water ranging from 500 ppm to 2%, significantly enhances nanotubes growth.

Since the lower limit of our oxygen mass flow rate is 1 sccm, we increased the H_2 flow rate of experiment E2 from 10 to 90 sccm in order to avoid the limit of explosion of the hydrogen/oxygen mixture. The conditions of this experiment called OPTI are summarized in Table 4.

In Figure 6 are showed the Scanning Electron Microscope images of sample OPTI. All the silicon surface is regularly covered by multiwalled carbon nanotubes of about 40 nm outer diameter and 1 **μm** long. Each nanotube is terminated by a catalyst particle on its top suggesting a top-growth mechanism.

In Figure 7, we can see the limit between the nickel covered and intentionally non covered witness area during the PVD step. Results demonstrate that the combination of TVA and PECVD is a powerful tool to uniform cover a large surface area with nanotubes.

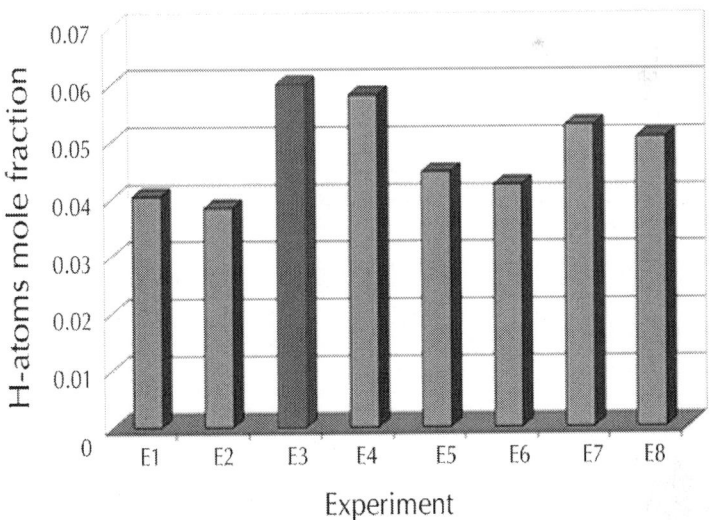

Figure 5: Calculated H-atom mole fraction for the 8 experimental conditions of Table 1.

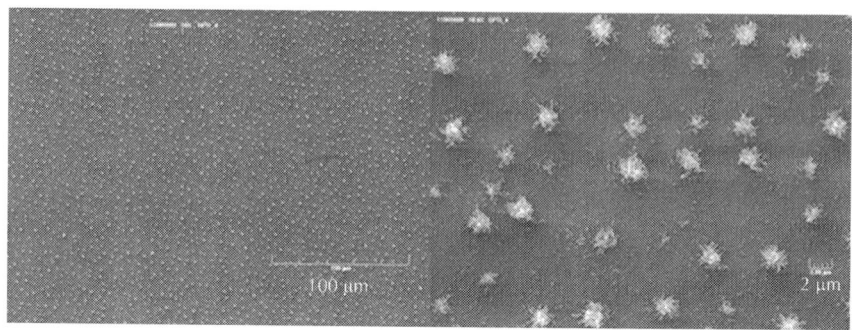

Figure 6: Scanning Electron Microscope images of samples OPTI showing a regular coverage of the substrate by nanotubes at two magnifications levels. Tool bars are100 **μm** for left image and 2 **μm** for right image.

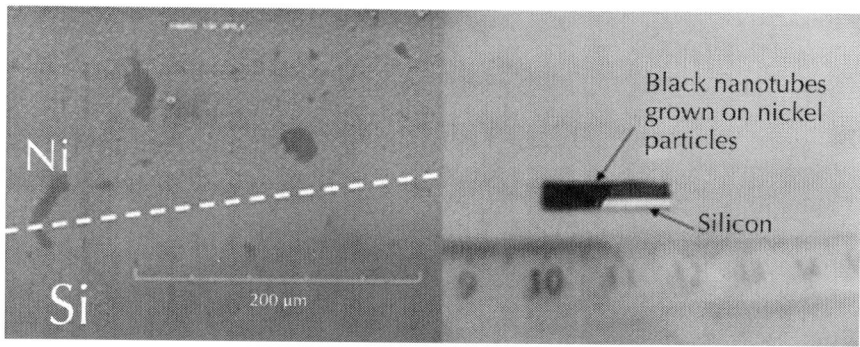

Figure 7: Left: Scanning Electron Microscope images of samples OPTI showing the border (dashed line) between nickel covered and non-covered silicon. Right: macroscopic image of the substrate, the black part is covered by nanotube and the shiny part is silicon.

Table 4: Optimal conditions, experiment OPTI

Experiment No.	Substrate Temperature Tsub (K)	Hydrogen Flow rate Q_{H_2} (sccm)	Methane Flow rate Q_{CH_4} (sccm)	Oxygen Flow rate Q_{O_2} (sccm)	Total pressure P (mbar)	Microwave power Pmw(Watt)
OPTI	1123	90	10	1	10	342.82

Modeling Results

A two dimensional steady-state reactor simulations performed by CFD code ANSYS Fluent 12 provided information about the temperature and species distribution in the reactor otherwise difficult to characterize. The simulations conditions are summarized in Table 5.

As first results we present the simulated temperature contours inside the reactor (Figure 8(a)). The maximal temperature in the center of the plasma at a distance of ~17 mm to the substrate is 2163 K. Such temperatures are expected to yield a complete dissociation of the carbon precursor and the availability of atomic carbon. Thermal balance of plasma heating and substrate cooling determines the substrate temperature. Hence, there is a steep temperature gradient between the substrate and the region where temperature is highest as shown in the 1D profile along the centerline of the reactor (Figure 8(b)). Also, temperature near the substrate is within the range for the appropriate MWCNT-synthesis temperature condition (950 - 1150 K). Thus, the experimentally observed, stably CNT-synthesizing conditions correspond to the appropriate conditions to produce MWCNTs, in terms of both supplied carbon concentrations and temperature.

The gas temperature decreases when the gas reaches around the quartz walls and the gas goes down.

The inlet gas is introduced in the reactor from the top side of the quartz enclosure. It splits into two components: one flows to the outlet and other flows to the substrate. Figure 9 shows the gas flow around the substrate. The flow and trajectories of gas species are visualized by using path lines. The length of arrows corresponds to the gas flow velocity. On the top surface of the substrate, gas flows from the center to the substrate edges. The gas stagnates and the magnitude of flow is relatively smaller than other regions. The presence of vortices in the region above the substrate is suspected to increase the mass flux of carbon species from the plasma zone to the cold substrate region, and to enhance the nanotube growth.

The simulated species profiles presented in Figure 10 show that C_2H_2, CH_3 and CH_4 are important species that may significantly contribute to carbon nanotubes growth. There is a region of uniform C_2H_2, CH_3 and CH_4 distribution where CNTs were synthesized for the experiments, showing a relatively broader region close to the substrate surface. Other species such as atomic carbon and hydrogen have also a rather important contribution. Finally, simulation shows that large amounts of H_2 are produced in the gas phase, but H_2 production arises also from the surface desorption (Figure 11).

First simulation results were obtained by adjusting the surface site density in order to reproduce the experimental deposition rate values. For   = 5.0 × 10^{-10} the calculated nanotube growth rate was 8.3 μm/h. We checked the influence of the substrate temperature by varying it from 873 K to 1273 K and confirmed that increasing the temperature leads to increasing of the nanotubes growth rate from 0.5 to 45 μm/h. To calculate the activation energy of the surface reactions, we plotted the logarithm of the calculated nanotubes growth rates against 10,000/ T_{sub} where T_{sub} is the temperature of the substrate (Figure 12). The plotted points follow an Arrhenius type law that allows us to calculate activation energy of 1.2 kJ/mol.

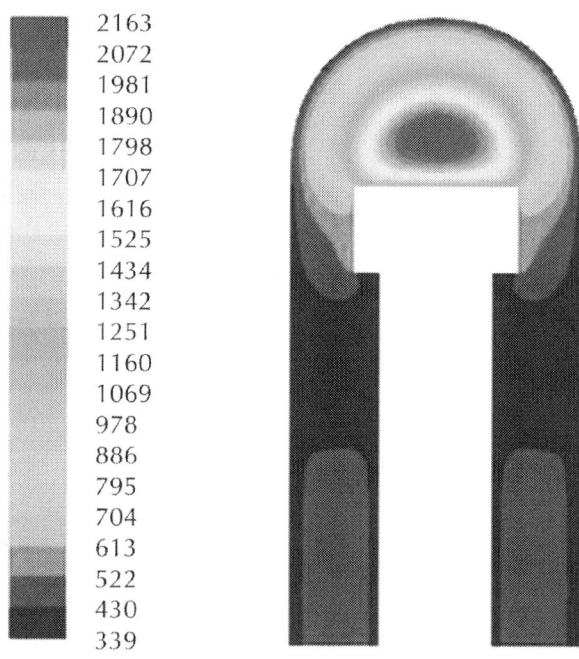

| 2163 |
| 2072 |
| 1981 |
| 1890 |
| 1798 |
| 1707 |
| 1616 |
| 1525 |
| 1434 |
| 1342 |
| 1251 |
| 1160 |
| 1069 |
| 978 |
| 886 |
| 795 |
| 704 |
| 613 |
| 522 |
| 430 |
| 339 |

Temperature, K

(a)

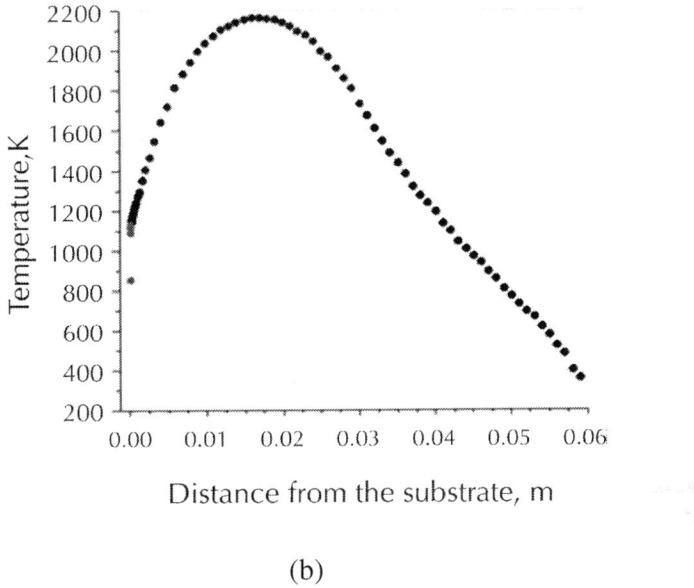

Distance from the substrate, m

(b)

Figure 8: (a) Simulated temperature profile inside the reactor; (b) 1D temperature profile along the centerline of the reactor.

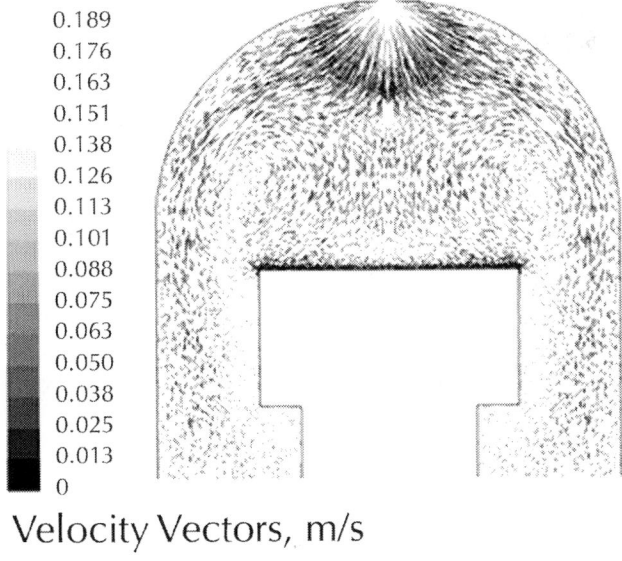

Velocity Vectors, m/s

Figure 9: Velocity vectors around the substrate.

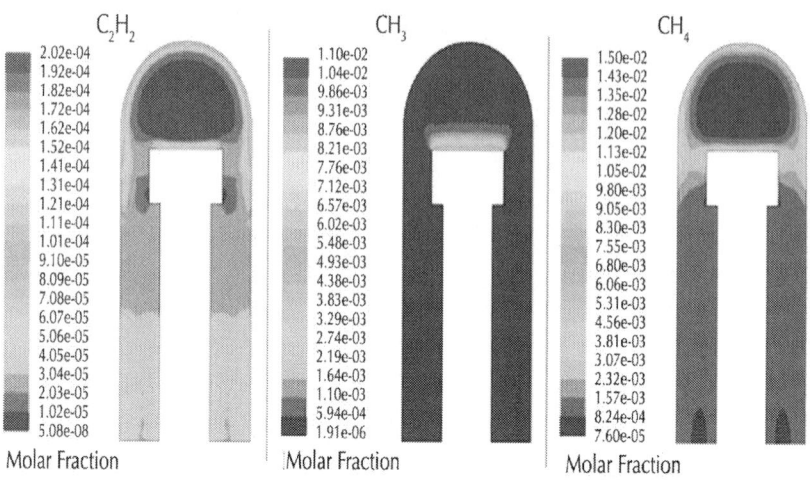

Figure 10: Simulated molar fractions of C_2H_2, CH_3 and CH_4.

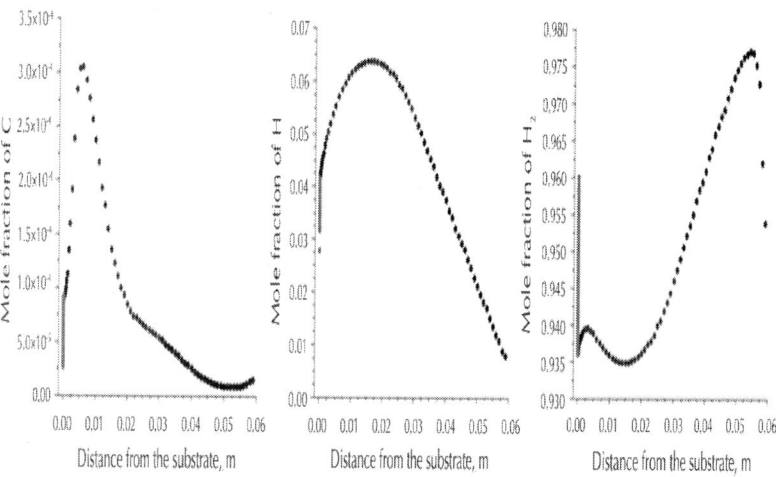

Figure 11: 1D temperature profile of the C, H and H_2 mole fractions along the centerline of the reactor.

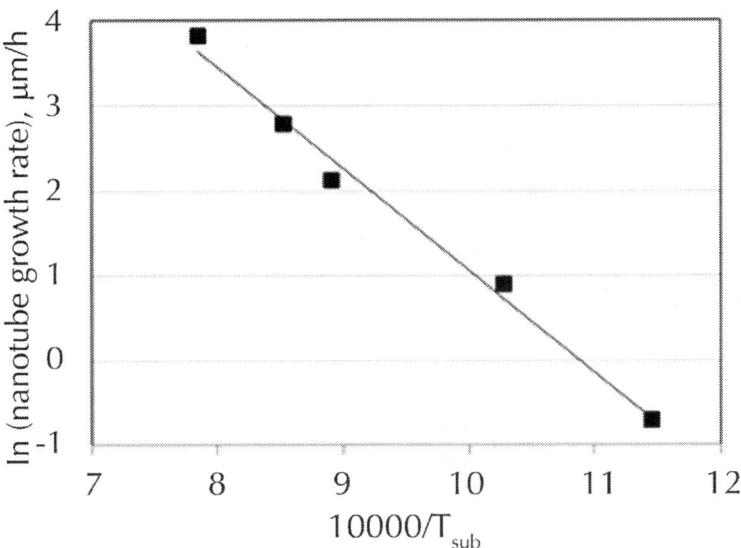

Figure 12: Logarithm of the calculated nanotubes growth rates against 10,000/T_{sub}.

Table 5: Simulations conditions

%vol. CH4	10
%vol. H2	90
Inlet Velocity, m/s	0.1326
Substrate Temperature, K	1123
Wall Temperature, K	400
Plasma Heat Source, W/m3	0.634×107
Pressure, mbar	10
Site density, mol/cm2	$5.0 \times 10{-}10$

CONCLUSIONS

In this work, we have successfully grown multi-walled carbon nanotubes on Ni/Si substrates using a combination of two methods: 1) thermionic vacuum arc (TVA) to catalyst 1 nm ultra-thin films deposition and 2) microwave plasma PECVD with a mixture of methane and hydrogen to CNT's growth. By using an experimental factor plan, substrate temperature and plasma power density were observed to significantly influence nanotube growth. Substrate temperature affects carbon diffusion into the catalyst particle while plasma power controls the atomic hydrogen in the plasma. Based on SEM observations, higher substrate temperature and lower hydrogen atom concentration are favorable to nanotube growth. In addition, a limited fraction of oxygen added to the plasma enhances the catalytic activity improving nanotube growth. Further work is underway in order to explore the number of walls and alignment of the CNTs by controlling the catalyst size.

Plasma reactor simulation results confirmed these experimental trends. Hydrocarbon species such as C_2H_2, CH_3 and C are likely to be key deposition species influencing CNT growth rate. The reaction mechanism used in these simulations will be improved to further confirm these preliminary results.

REFERENCES

1.   Iijima, S. (1991) Helical Microtubules of Graphitic Carbon. Nature, 354, 56-58.http://dx.doi.org/10.1038/354056a0

2.   Farhat, S. and Scott, C. (2006) Review of the Arc Process Modeling for Fullerene and Nanotube Production. Journal of Nanoscience and Nanotechnology, 6, 1189-1210.http://dx.doi.org/10.1166/jnn.2006.331

3.   Yu, B. and Meyyappan, M. (2006) Nanotechnology: Role in Emerging Nanoelectronics. Solid State Electronics, 50, 536-544. http://dx.doi.org/10.1016/j.sse.2006.03.028

4.   Schäffel, F., Schünemann, C., Rümmeli, M.H., Täschner, C., Pohl, D., Kramberger, C., Gemming, T., Leonhardt, A., Pichler, T., Rellinghaus, B., Büchner, B. and Schultz, L. (2008) Comparative Study on Thermal and Plasma Enhanced CVD Grown Carbon Nanotubes from Gas Phase Prepared Elemental and Binary Catalyst Particles. Physica Status Solidi (b), 245, 1919-1922. http://dx.doi.org/10.1002/pssb.200879605

5.   Delzeit, L., Nguyen, C.V., Stevens, R.M., Han, J. and Meyyappan, M. (2002) Growth of Carbon Nanotubes by Thermal and Plasma Chemical Vapour Deposition Processes and Applications in Microscopy. Nanotechnology, 13, 280-284.http://dx.doi.org/10.1088/0957-4484/13/3/308

6.   Hata, K., Futaba, D.N., Mizuno, K., Namai, T., Yumura M. and Ijima, S. (2004) Water-Assisted Highly Efficient Synthesis of Impurity-Free Single-Walled Carbon Nanotubes. Science, 306, 1362-1364. http://dx.doi.org/10.1126/science.1104962

7.   Zhang, G., Mann, D., Zhang, L., Javey, A., Li, Y., Yenilmez, E., Wang, Q., McVittie, J.P., Nishi, Y., Gibbons, J. and Dai, H. (2005) Ultra-High-Yield Growth of Vertical Single-Walled Carbon Nanotubes: Hidden Roles of Hydrogen and Oxygen. PNAS, 102, 16141-16145.http://dx.doi.org/10.1073/pnas.0507064102

8.   Lungu, C.P., Mustata, I., Zaroschi, V., Lungu, A.M., Anghel, A., Chiru, P., Rubel, M., Coad, P. and Matthews, G.F. (2007) Beryllium Coatings on Metals for Marker Tiles at JET: Development of Process and Characterization of Layers. Physyca Scripta, T128, 157-161.http://dx.doi.org/10.1088/0031-8949/2007/T128/030

9.   Lungu, C.P., Mustata, I., Musa, G., Lungu, A.M., Zaroschi, V., Iwasaki, K., Tanaka, R., Matsumura, Y., Iwanaga, I., Tanaka, H., Oi, T. and Fujita, K. (2005) Formation of Nanostructured Re–Cr–Ni Diffusion Barrier Coatings on Nb Superalloys by TVA Method. Surface and Coating Technology, 200, 399-402.http://dx.doi.org/10.1016/j.surfcoat.2005.02.172

10. Lungu, C.P. (2005) Nanostructure Influence on DLC-Ag Tribological Coatings. Surface and Coating Technology, 200, 198-202. http://dx.doi.org/10.1016/j.surfcoat.2005.02.103

11. Lungu, C.P., Mustata, I., Zaroschi, V., Lungu, A.M., Chiru, P., Anghel, A., Burcea, G., Bailescu, V., Dinuta, G. and Din, F. (2007) Spectroscopic Study of Beryllium Plasma Produced by Thermionic Vacuum Arc. Journal of Optoelectronics and Advanced Materials, 9, 884-886.

12. Silva, F., Gicquel, A., Chiron, A. and Achard, J. (2000) Low Roughness Diamond Films Produced at Temperatures Less than 600°C. Diamond and Related Materials, 9, 1965-1970. http://dx.doi.org/10.1016/S0925-9635(00)00347-2

13. Scott, C.D., Farhat, S., Gicquel, A., Hassouni, K. and Lefebvre, M. (1996) Determining Electron Temperature and Density in a Hydrogen Microwave Plasma. Journal of Thermophysics and Heat Transfer, 10, 426-435. http://dx.doi.org/10.2514/3.807

14. Garg, R.K., Kim, S.S., Hash, D.B., Gore, J.P. and Fisher, T. (2008) Effects of Feed Gas Composition and Catalyst Thickness on Carbon Nanotube and Nanofiber Synthesis by Plasma Enhanced Chemical Vapor Deposition. Journal of Nanoscience and Nanotechnology, 8, 3068-3076. http://dx.doi.org/10.1166/jnn.2008.082

15. Marinov, N.M. and Malte, P.C. (1995) Ethylene Oxidation in a Well-Stirred Reactor. International Journal of Chemical Kinetics, 27, 957-986.http://dx.doi.org/10.1002/kin.550271003

16. Walter, D., Grotheer, H.H., Davies, J.W., Pilling, M.J. and Wagner, A.F. (1990) Experimental and Theoretical Study of the Recombination Reaction CH3 + CH3 – C2H6. Symposium (International) on Combustion, 23, 107-114.http://dx.doi.org/10.1016/S0082-0784(06)80248-1

17. Tsang, W. and Hampson, R.F. (1986) Chemical Kinetic Data Base for Combustion Chemistry. Part 1. Methane and Related Compounds. Journal of Physical and Chemical Reference Data, 15, 1087-1279. http://dx.doi.org/10.1063/1.555759

18. Miller, J.A. and Melius, C.F. (1992) Kinetics and Thermodynamic Issues in the Formation of Aromatic Compounds in Flames of Aliphatic Fuels. Combustion and Flame, 91, 21-39.http://dx.doi.org/10.1016/0010-2180(92)90124-8

19. Markus, M.W., Woiki, D. and Roth, P. (1992) Two-Channel Thermal Decomposition of CH3. Symposium (International) on Combustion, 24, 581-588.http://dx.doi.org/10.1016/S0082-0784(06)80071-8

20. Warnatz, J. (1984) Rate Coefficients in the C/H/O System. In: Gardiner Jr., W.C., Ed., Combustion Chemistry, Book Chapter, and Springer-Verlag, New York.

21. Dagaut, P., Cathonnet, M., Aboussi, B. and Boettner, J.-C. (1990) Allene Oxidation: A Kinetic Modeling Study. Journal de Chimie Physique ET de Physico-Chimie Biologique, 87, 1159-1172.

22. Feng, Y., Niiranen, J.T., Bencsura, A., Knyazev, V.D., Gutman, D. and Tsang, W. (1993) Weak Collision Effects in the Reaction C2H5=C2H4+H. Journal of Physical Chemistry, 97, 871-880. http://dx.doi.org/10.1021/j100106a012

23. Towell, G.D. and Martin, J.J. (1961) Kinetic Data from Nonisothermal Experiments: Thermal Decomposition of Ethane, Ethylene, and Acetylene. AIChE Journal, 7, 693-698. http://dx.doi.org/10.1002/aic.690070432

24. Kiefer, J.H., Kapsalis, S.A., MAlami, M.Z. and Budach, K.A. (1983) the Very High Temperature Pyrolysis of Ethylene and the Subsequent Reactions of Product Acetylene. Combustion and Flame, 51, 79-93. http://dx.doi.org/10.1016/0010-2180(83)90085-8

25. Dean, A.M. (1985) Predictions of Pressure and Temperature Effects upon Radical Addition and Recombination Reactions. Journal of Physical Chemistry, 89, 4600-4608. http://dx.doi.org/10.1021/j100267a038

26. Fahr, A., Laufer, A., Klein, R. and Braun, W. (1991) Reaction Rate Determinations of Vinyl Radical Reactions with Vinyl, Methyl, and Hydrogen Atoms. Journal of Physical Chemistry, 95, 3218-3224. http://dx.doi.org/10.1021/j100161a047

27. Knyazev, V.D., Bencsura, A., Stoliarov, S.I. and Slagle, I.R. (1996) Kinetics of the C2H3+H2=H+C2H4 and CH3+ H2=H+CH4 Reactions. Journal of Physical Chemistry, 100, 11346-11354. http://dx.doi.org/10.1021/jp9606568

28. Janardhanan, V.M. and Deutschmann, O. (2006) CFD Analysis of a Solid Oxide Fuel Cell with Internal Reforming: Coupled Interactions of Transport, Heterogeneous Catalysis and Electrochemical Processes. Journal of Power Sources, 162, 1192-1202. http://dx.doi.org/10.1016/j.jpowsour.2006.08.017

29. Lysaght, A.C. and Chiu, W.K.S. (2008) Modeling of the Carbon Nanotube Chemical Vapor Deposition Process Using Methane and Acetylene Precursor Gases. Nanotechnology 19, 165607. http://dx.doi.org/10.1088/0957-4484/19/16/165607

30. Lacroix, R., Fournet, R., Ziegler-Devin, I. and Marquaire, P.-M. (2010) Kinetic Modeling of Surface Reactions Involved in CVI of Pyrocarbon Obtained by Propane Pyrolysis. Carbon, 48, 132–144. http://dx.doi.org/10.1016/j.carbon.2009.08.041

31. Farhat, S., Panham, S., Gicquel, A., Silva, F., Brinza, O. and Lungu, C.P. (2010) Synthèse de Nanotubes Orientés par PECVD. Matériaux 2010, 18-22 October 2010, Nantes.

32. Kee, R.J., Rupley, F.M., Miller, J.A., Coltrin, M.E., et al. (2001) CHEMKIN Collection, Release 3.6, Reaction Design, Inc., San Diego.

33. JANAF (1965) "Thermochemical tables, National Standards Reference Data Series" Report NSRDS-NBS: Dow Chemikal Company, distributed by Clearinghouse for federal Scientific and Technical Information, PB168370.

34. Kee, R.J., Dixon-Lewis, G., Warnatz, J. and Miller, J.A. (1986) A FORTRAN Computer Code Package for Evaluation of Gas-Phase, Multicomponent Transport Properties. Technical Report SAND86-8426, Sandia National Laboratories, Albuquerque.

35. Sickafus, E.N. (1970) Sulfur and Carbon on the (110) Surface of Nickel. Surface Science, 19, 181-197. http://dx.doi.org/10.1016/0039-6028(70)90117-2

36. Kee, R.J., Miller, J.A. and Jefferson, T.H. (1980) CHEMKIN: A General-Purpose, Problem-Independent, Transportable, Fortran Chemical Kinetics Code Package. Technical Report SAND80-8003, Sandia National Laboratories, Albuquerque.

37. Farhat, S., Findeling, C., Silva, F., Hassouni, K. and Gicquel, A. (1997) Third Edition of the International Workshop Microwave Discharges: Fundamentals and Applications. Abbaye de Fontevraud, Fontevraud-l'Abbaye.

Chapter 8

Studies on Chemical Resistance of PET-Mortar Composites: Microstructure and Phase Composition Changes

Yassine Senhadji[2], Ilies Bahlouli[1], and David Houivet[4]

[1]Faculty of Science, Laboratory of Polymer Chemistry, University of Oran, Es-Senia, Algeria

[2]Department of Civil Engineering, Laboratory of Materials, ENSET, Oran, Algeria

[3]Department of Chemistry, Preparatory School of Science and Technology EPST, Tlemcen, Algeria

[4]University of Caen Basse-Normandie, Laboratory of LUSAC EA 2607, Cherbourg Octeville, France

ABSTRACT

Researches into new and innovative uses of waste plastic materials are continuously advancing. These research efforts try to match society's need for safe and economic disposal of waste materials. The use of recycled plastic aggregates saves natural resources and dumping spaces, and helps to maintain a clean environment. The present articles deals with the resistance to chemical attack of polymer-mortars, which are often used as low-cost promising materials for preventing or repairing various reinforced concrete structures. To gain more knowledge on the efficiency of polymer-mortar composites, four mortar mixtures: one specimen with Portland cement and three mixtures with 2.5, 5, and 7.5 wt% of the substitution of cement by polyethylene terephthalate (PET) were exposed to the influence of aggressive environment (0.5%, 1% and 1.5% HCl acids, 10% NH_4Cl, 5% H_2SO_4 acid and 10% $(NH_4)_2SO_4$ solutions). The measurements of several properties were carried out, the results were analyzed and the combination of X-ray diffraction, FT-IR spectroscopy, differential thermal analysis (DTA), thermo gravimetric (TG) analysis, differential scanning calorimetry (DSC) analysis and the composites were also observed by SEM led to the positive identification of the deterioration products' formation. From this study, it was found that the addition of PET to the modified mortars, means reducing the penetration of aggressive agents. So, the PET-modified mortars exposed to aggressive environments showed better resistance to chemical attack. The new composites appear to offer an attractive low-cost material with consistent properties. The present study highlights the capabilities of the different methods for the analysis of composites and opened new way for the recycling of PET in polymer-mortars.

INTRODUCTION

Degradation of concrete, as well as the related protection and ensuring of concrete structures against aggressive impacts by chemical agents, regardless of whether this concerns liquid, gas, or even solid phase under certain conditions, represents a complex problem of utmost importance for the economy in general, and especially for building construction and the construction industry [1]. Therefore, polymer-modified mortars have been popular construction materials because of their excellent properties in comparison with ordinary mortars. Polymers have been used for improving mechanical properties, adhesion with substrates, or waterproofing properties of mortars and concretes. The literature agrees that the properties of polymer modified mortar and concrete depend significantly on the polymer content or polymer-cement ratio, that is, the mass ratio of the amount of polymer solids in a polymer-based admixture to the amount of cement in a polymer-modified mortar or concrete [2-4].

A substantial growth in the consumption of plastic is observed all over the world in recent years, which has led to huge quantities of plastic-related waste. Recycling of plastic waste to produce new materials like concrete, mortar or composite appears as one of the best solution for disposing of plastic waste, due to its economic and ecological advantages. Several works have been performed or are under way to evaluate the properties of cement-composites containing various types of plastic waste as aggregate, filler or fibre [5,6].

Depending on the appropriate final target, varied types of waste can be used in concrete: Portella et al. [7] and Guerra et al. [8] have studied the properties of concrete with addition of ceramic waste; Ismail and Al-Hashmi [9] have studied concrete with recycled plastic addition; Angulo et al. [10] have characterized recycled materials from construction and demolition as an aggregate for concrete; and, amongst others, Hoppen et al. [11] have found in concrete the possibility to deposit the residual sludge from water treatment stations.

Different works, as of Rebeiz [12], Choi et al. [13,14] and JO et al. [15], have analyzed the effect of addition of recycled PET to the properties of concrete. The fibers of recycled PET easily mix in the concrete, giving new properties to the material [16]. Khaloo et al. [17] have observed that the addition of tire rubber particles provided the concrete with higher ductility in compressive strength testing, if compared with concrete without addition. J. C. A. Galvão et al. [18] have demonstrated a better performance to use of waste polymers (PET, LDPE and rubber from useless tires) in concrete for repair of dam hydraulic surfaces. Wang et al. [19] have analyzed a Performance of cement mortar made with recycled high impact polystyrene, which is a common component of consumer electronics. Corinaldesi et al. [20] have also showed a better mechanical behaviour and thermal conductivity of mortars containing waste rubber particles coming from wasted rubber-shoe outsoles (SR, acronym of "sole rubber"). In the previous work [21,22], the author studied the effects of PET polymer on the mortar properties, specifically to decrease the chloride ion penetration depth and apparent chloride ion diffusion coefficient of polymer-mortar composites. This may be explained due to the reduced volume of large-sized pores and the improved resistance to the absorption of the test solutions with an increase in polymer-cement ratio [22]. In addition, Gouasmi et al. [23,24] showed the improvement of the adherence strength and the resistance to aggressive solutions of composites using waste PET lightweight aggregates (WPLA). One of the advantages of the use of recycled plastic in concrete is the reduction of solid waste in landfills [5,6].

Polyethylene terephthalate (PET) is one of the most common consumer plastics used and is widely employed as a raw material to realize products such as blown bottles for soft-drink use and containers for the packaging of food and other consumer goods. PET bottles have taken the place of glass bottles as storing vessel of beverage due to its lightweight and easiness of handling and storage.

In 2007, it is reported a world's annual consumption of PET drink covers of approximately 10 million tons, which presents perhaps 250 milliards bottles. This number grows about up to 15% every year [25]. On the other hand, the number of recycled or returned bottles is very low. Generally, the empty PET packaging is discarded by the consumer after use and becomes PET waste (WPET). The major problems that this level of waste production generates initially entail storage and elimination [26]. The recycling of PET bottles and the preservation of natural resources are priority items but to date, the recycling of PET bottles as a lightweight aggregate for concrete has not been studied because of the high melting cost [16].

Chemical degradation of concrete is the consequence of reactions between the constituents of cement stone, i.e., calcium silicates, calcium aluminates, and above all calcium hydroxide, as well as other constituents, with certain substances from water, solutions of soil, gases, vapors, acids etc. The most important aggressive agents are: $SO_4^{2-}, Mg^{2+}, NH_4^+, Cl^-, H^+$, and HCO_3^- [1,27,28].

When speak about sulfate degradation, we primarily think of the impact by sulfate ions on cement stone. The sulfate ion is the cause of one of the most dangerous corrosions—the corrosion of expansion and swelling—because it causes the occurrence of expansive compounds, the most important of which is ettringite, $C_3A \cdot 3CaSO_4 \cdot 32H_2O$, in the shape of prismatic crystals [29,30].

For the process of concrete degradation under the impact of sulfates, it is essential which cation is linked with the sulfate ion. Namely, cations linked with sulfate ions can be divided into three characteristic groups [1]. The first group includes alkali metals Na^+ and K^+, which give extremely soluble hydroxides, while the second group comprises metals such as Mg^{2+} and Fe^{2+}, which give poorly soluble hydroxides, and the third group consists of cations NH_4^+ and H^+, which give volatiles or hydroxide. The third group of sulfates, that is $(NH_4)_2SO_4$ and H_2SO_4, covers the most aggressive compounds. In case of impact by these compounds on concrete, there occurs not only expansion, but also intensive dissolution of cement stone [1].

The degree of aggressivity of an acid is dependent on the chemical character of anions present. The strength of acid, its dissociation degree in solutions and, mainly, the solubility of the salt formed are dependent on the chemical character of anion. With respect to concrete acidic attack, the solubility of calcium salts formed is of a great significance. Because the acidic attack is based on the processes of decomposition and leaching of the constituents of cement matrix, the conditions of transport phenomena, such as the supply of aggressive acidic solutions and draining of the products of the attack is of specific significance for the intensity of attack [31-33].

In addition, a parameter which is tightly connected with the properties of acids is that their pH reaches a value of approximately 5 and lower. The severity of the acidic attack is significantly dependent on the solubility of the calcium salt formed. In the case of the formation of highly soluble salts, the severity of the attack is very high. This is caused by the dissolving and leaching of the formed salt from the attacked material. A very porous layer of corrosive products remains on the surface, contributing to further development of the deterioration process. In the case of the formation of insoluble calcium salts such as calcium oxalate and sulfate the effect of the acidic solution is entirely different. A dense insoluble layer is formed enhancing the development of the acidic attack showing a protective effect. This is used in practice for the protective purposes [34].

The objective of the research reported here was to evaluate the chemical resistance of polymer-mortar composites containing a waste polyethylene terephtalate PET, as a substitute of the cement used in the mix, under hydrochloric acid, ammonium-chloride, ammonium-sulfate and sulfuric acid solutions exposure. The identification of the deterioration products' which appear on the surface of the samples were analyzed by X-ray diffraction, FTIR, SEM, ATD, TG/dTG and DSC analyses. However, no pertinent data were previously found concerning the effect of PET against ammonium-chloride or ammoniumsulfate attack. So, in this paper the durability of PETmortar composites exposed to aggressive solutions has been investigated. If successful, such an application of waste PET, as substitute of cement, will be a major step towards reducing the solid waste disposal problem and reliance on natural resources, thereby reducing environmental pollution and energy consumption. Additionally, the immediate consequence is the anticipated necessity of maintenance and repairs of the concrete structures, which must have specific characteristics, mainly mechanical and chemical, based on the material of the base or its substratum.

MATERIALS, EXPERIMENTAL DESIGN AND METHODS

Raw Materials

The cement used was a blended Portland cement type CPJ-CEM II/A (pouzzolanic cement) delivered from Zahana factory located in the western Algeria, chemical and physical properties of cement are shown in Tables 1 and 2, respectively, according to the manufactories. The chemical composition was obtained by using an X-ray fluorescence spectrometer analysis type OXFORD MD X^{1000}.

Two different aggregate types were selected for this study, sand (S) and PET particles. Those come from drinking water bottles that was first separated, washed and shredded. The particles thus derived were then shredded once again, using a propeller crusher in order to control granular limit with crushing. Moreover, they present irregular shape and rough surface texture in order to facilitate matrix-particles adhesion (Figure 1). Figure 2 showed the X-ray diffraction analysis of PET particles and indicated the presence of a backbone form of the polymer which generates sharp peaks at a high angle range ~10° - 35°. The main mechanical and thermal properties of the plastic used in this study are presented in Table 3. Particle size distributions analysis (Figure 3) reveals the sand to be particularly coarse (Fineness modules = 3.66) and that the upper granular limit value of PET aggregates is 5 mm. After preliminary tests, polymer particles of size lower than 1 mm

were used in this study. Absolute density of PET particles $\left(1.35g/cm^3\right)$

is approximately 2.5 times smaller than that of cement $\left(3.09g/cm^3\right)$. This low density will allow the lightening of PET-mortar composite.

Table 1: Cement chemical composition

Oxide analysis (%)					
Loss on ignition	2.08	Fe_2O_3	4.57	SO_3	1.85
SiO_2	21.83	C_aO	63.42	C_aO free	0.25
Al_2O_3	6.01	M_gO	0.22		

Table 2: Physical properties of cement

Physical properties	
Setting time (min)	
Initial	120
Final	200
Blaine surface area (cmig)	2977
Absolute density (g/cm³)	3.09
Compressive strength (MPa) 28 days	32.5

Table 3: Physical properties of the PET plastic used

Physical properties	
Tensile strength at break (MPa)	70
Elongation at break (%)	70
Flexural modulus (rigidity) (MPa)	2.0
Tensile modulus (GPa)	2.9
Melting point (°C)	260
Water absorption (%)	0

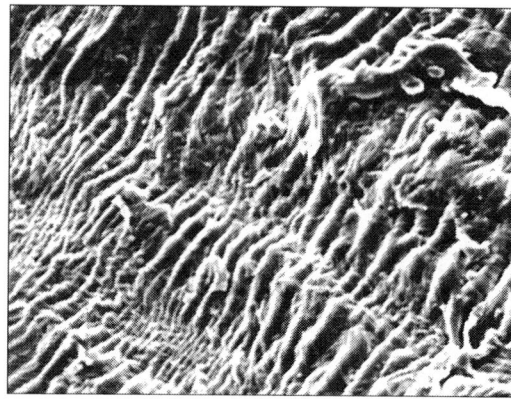

Figure 1: Scanning electron micrographs of PET particles, (magnification, ×1200).

Composite Mixing Conditions

The mortar manufactured with PET particles was first optimized on the basis of mechanical criteria and then constitutes the reference composite. The composites containing PET particles were produced in accordance with the results of the previous work [35]. A massic ratio of 3 between sand (S) and the cement (C) has been respected. Various massic percentages of cement (2.5%, 5.0% and 7.5%) were substituted by the same weight of granulated plastic waste (Table 4). The water to binder ratio was kept constant at 0.5. The physical properties of the pastes of mortars were determined in accordance with EN 19-3 [36].

The PET particles and cement were first dry-mixed at 140 rpm during 1 min in a standard mixing machine (EN 196-1 [37]) to reach a homogeneous mixture. Then sand was added and the mixing continues during 1 min at low speed. Next, water was gradually added at 140 rpm during 1 min and after the whole is mixed during 1 min with 285 rpm. After pouring fresh material into the molds, samples were stored in both a hygrometrically-controlled and temperature-controlled room (98% relative humidity and 20°C ± 2°C) for 24 h. After removal from the moulds, at 24 h of age, mortar specimens were immersed in water saturated with lime at 20°C ± 3°C until the age of testing. Test results from the Table 4 of the hydraulic transport properties, sorptivity-value, revealed that the addition of PET particles tends to restrict water propagation in the cement matrix and reduces water absorption of the composite. This may be due both to the capability of PET to repel water (non-sorptive nature, water absorption % = 0, Table 3) and to the increase of air-entrainment, as manifested by closed empty pores, which are not accessible to water. This phenomenon serves to reduce the volume accessible to water and hence capillary porosity. The decrease in water absorption is also attributed to a reduction in the porosity near particle/matrix interfacial zone, due to the high bonding between PET additive and cement paste [35]. So, the decrease of the sorptivity-value is favorable to the durability of the specimen structures.

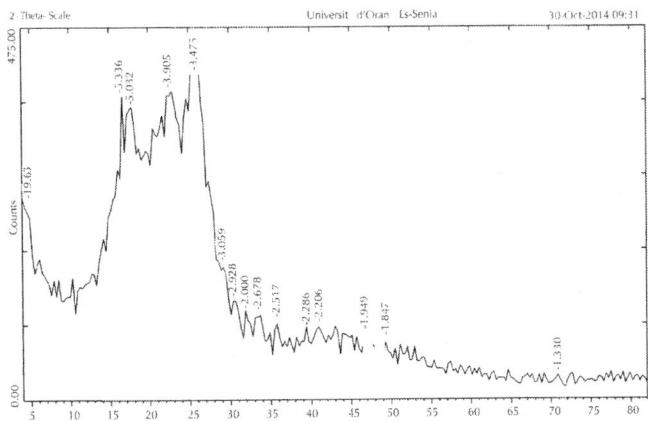

Figure 2: X-ray diffraction pattern of the PET particles.

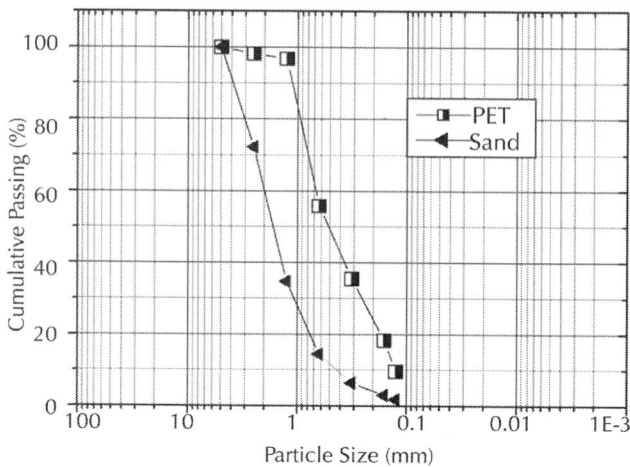

Figure 3: Particle size distributions of polyethylene terephthalate (PET) and Sand.

Resistance to Chemical Attack Test

The relative acid attack was determined in accordance with ASTM C267-97 [38]. The mortar specimens were cured in water at 20°C ± 3°C for 28 days before being subjected to acid attack. Three specimens of

each mortar and composite mixes $(50 \times 50 \times 50 \text{mm}^3)$ were immersed in three types of chemical solutions: 10% ammonium chloride NH_4Cl; 10% ammonium-sulfate $(NH_4)_2SO_4$; 5% sulfuric acid H_2SO_4. Before the test, the attacked specimens were cleaned with deionised water and then the acid attack was evaluated by measuring the mass loss (ML) of the specimens, determined as follows:

$$ML\left(\%\right) = \frac{W_r - W_s}{W_r} \times 100$$

(1)

where W_r is weight of the specimen before immersion and W_s is weight of the cleaned immersed specimen after test period. The solution was renewed every 7 days and the mass loss of the specimens measured.

After immersion in 0.5%, 1% and 1.5% HCl acids $\left(4\times4\times16\text{mm}^3\right)$, 5% H_2SO_4 acid (ASTM C267-97) and 10% $(NH_4)_2SO_4$ solutions (ASTM C1012-04 [39]) for the required period of time, the specimens were capped and tested for residual compressive strength based on the original cross-sectional area. The compressive strength loss (CSL%) is calculated as follows:

$$CSL\left(\%\right) = \frac{f_{cr} - f_{cs}}{f_{cr}} \times 100$$

(2)

where f_{cr} is the reference compressive strength of specimen before immersion in the acid or sulphate solutions in MPa and f_{cs} is the average compressive strength of the specimens after immersion in acid or sulphate solutions for the required period of time.

After compression testing, scanning electron microscopy (SEM), X-ray diffraction (XRD), FT-IR analyses, DSC and DTA-TGA/dTG were conducted on selected surface fractures to investigate damage mechanisms.

An accelerated leaching test using an ammonium chloride solution (10% NH_4Cl) was carried out on three samples $\left(50\times50\times50\text{mm}^3\right)$ of each composite, after cured in water at 20°C ± 3°C for 28 days. Ammonium nitrate allows an equivalent leaching test with demineralised water to be simulated but increases the kinetics by a factor 100 [40,41]. The mortar and composites specimens were immersed in the solution at 20°C for 480 days. For the required time of test, the leached depth (observed with phenolphthalein) and the mass loss were measured each time of test.

RESULTS AND DISCUSSION

Hydrochloric Acid (HCl) Attack

Compressive Strength Loss (CSL%)

Mass loss is a simple traditional test in the context of acid attack. However, mass change results may depend on sample size and cement type, and are also influenced by the way the decomposed cement paste and other reaction products on samples are treated during testing [42, 43]. Therefore, along with mass loss test, compressive strength is considered to be a more reliable measure to judge the performance of mortar/concrete subjected to acid attack. Siad et al. [44] reported that there is some divergence between the mass loss and the compressive strength loss.

Table 4: Mix proportions and physical properties of polymer-mortar composites

Polymer-Cement Ratios (%)	Setting Time (min)		Water demand for Standard Consistency (%)	Density (g/cm3)	Compressive strength (MPa)	Capillary Coefficient E−05 (cm2/sec)
	Initial	Final			28 days	28 days
0	120	200	24	2.28	41.6	3.89
2.5	125	205	24.5	2.23	38.6	3.50
5.0	130	210	25	2.22	36.6	3.13
7.5	145	225	25.5	2.21	32.5	2.85

Hence, Figure 4 presents the results of compressive strength loss CSL% of the mixes at 56 days in different hydrochloric acid concentration (0.5%, 1% and 1.5%) solutions. The results indicate that the resistance of acidic attack of the composites (PET7.5) was increased with an increase in PET content. At day-56, the CSL% of PET7.5 was reduced by 68%, 41%, 10.7% for the 0.5%, 1% and 1.5% of HCl acid solutions, respectively, when compared to that of PET0.

The H2SO4 solutions lead to the formation of less water-soluble gypsum CaSO4, $0241g / 100ml$ of H2O at 20°C, and ettringite on the surface in contact with the cementing matrix. With CH3COOH acid, there is a formation of the hydrate calcium acetate Ca(CH3COO)2, $52.0g / 100ml$ of H2O. While, the chemicals formed as the products of reaction between hydrochloric acid and hydrated cement phases are some soluble salts, mostly with calcium chloride CaCl2 (dihydrate) that is very watersoluble: $46.08g / 100ml$ of H2O, which are subsequently leached out, and some insoluble salts along with amorphous hydrogels (iron hydroxide) which remain in the corroded layer. These chemical reactions are shown in Equations (3)-(5). These results are in agreement with those reported by elsewhere [45-47]

$$2HCl + Ca(OH)_2$$
$$\rightarrow CaCl_2 \cdot 2H_2O\,(\text{hydrate calcium chloride}); \tag{3}$$

$$H_2SO_4 + Ca(OH)_2 \rightarrow CaSO_4 \cdot 2H_2O\,(\text{gypsum}); \tag{4}$$

$$3CaSO_4 + 3CaO \cdot Al_2O_3 \cdot 6H_2O + 25H_2O$$
$$\rightarrow 3CaO \cdot Al_2O_3 \cdot 3CaSO_4 \cdot 31H_2O\,(\text{ettringite}) \tag{5}$$

Additionally, these results are confirmed by the change of surface samples before and after immersion in the HCl aggressive solutions as depicted in Figure 5. These mortars kept their rectangular forms more or less, but their dimensions decreased considerably from 0.5%, 1% to 1.5%. Hydrochloric acid attack is a typical acidic corrosion which can be characterized by the formation of layer structure [47]; its can be divided to three main zones: undamaged zone, hydroxide mixture zone or brown ring, and attacked zone. By hydroxide mixture zone, there is a layer formed by undissolved salts seen as a dark brown ring.

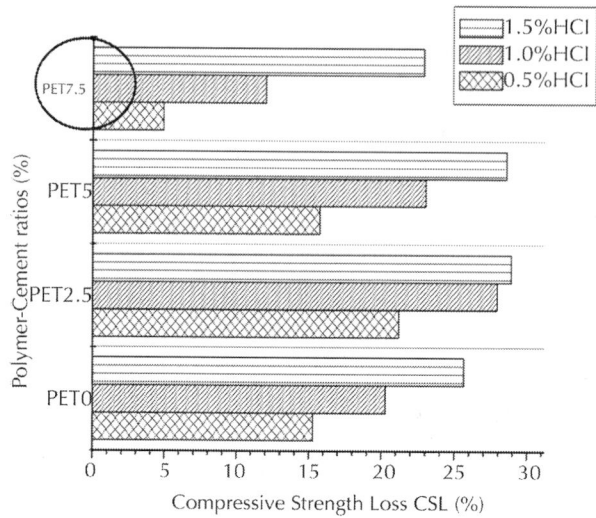

Figure 4: Compressive strength loss of specimens under HCl acid solutions exposure.

The chemical resistance of materials is more or less affected by the concentration and the nature of acids in the order with the most aggressive as given below:

$$0.5\%HCl < 1.0\%HCl < 1.5\%HCl \ .$$

The increase in the resistance to hydrochloric attack of the composites is attributed to the impervious PET granules blocking the passage of the aggressive solutions and the reduction of the sorptivity of PET-mortar composites (Tables 3 and 4). Furthermore, the decrease in porosity due to the incorporation of PET in modified mortars [48] contributes to reduce the absorption of acidic solution accompanied by a reduction of loss in weight. These results are in agreement with those reported by Benosman et al. [46]. Additionally, different teams of researchers [49-51] reported that the incorporation of organic additions (polymers) increases chemical resistance in aggressive media.

X-Ray Diffraction (XRD) Analysis

Figure 6 presents the XRD analysis of composite PET0, as an example, before and after attack by different hydrochloric acid concentration

(0.5%, 1% and 1.5%) solu tions. The stacking of the various spectra (initial state, after 0.5% attack, after 1% attack and after 1.5% attack) confirms the appearance of a trace of calcium chloride ($CaCl_2$) on the specimens exposed to hydrochloric acid. The low quantity of calcium chloride found is due to its high solubility in water (46.08g / 100ml of H_2O). Washing of each specimen after the immersion period leaves its surface nearly free from this salt. Additionally, the portlandite $Ca(OH)_2$ was decomposed by different acids concentration following the chemical reaction (Equation (3)).

Figure 5: Deterioration of specimens after 56 days of immersion; (1) Tap water, (2) 0.5%, (3) 1%, (4) 1.5% HCl acids (From the left to right, respectively).

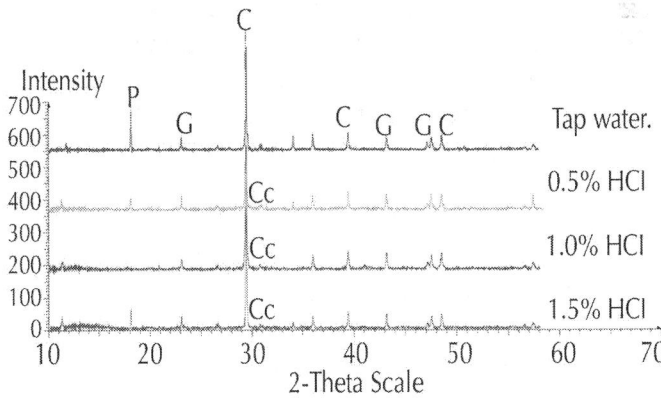

Figure 6: X-ray diffraction pattern of the specimens under HCl acids exposure. P: portlandite Ca $(OH)_2$, C: calcite, G: gypsum, Cc:$CaCl_2.2H_2O$ (hydrated calcium chloride).

FT-IR Analyses

Table 5 illustrates the positions and intensities of infrared absorption bands and Figure 7 shows the FT-IR patterns of specimens exposed to: Tap water, 0.5%, 1% and 1.5% HCl acids solution. The FT-IR spectra of the composite hydrated up to 56 days and cured in water are presented inFigure 7 and Table 5. The major changes of the FT-IR spectra in the hydrated cement pastes are: Calcium hydroxide bands (~ 3635 cm^{-1}) and also for the free OH groups, combined and adsorbed water of CSH, AFm and AFt phases (~ 3480 cm^{-1}), molecular water (3440 - 3446 and 1614 - 1621 cm^{-1}), carbonate phases (~ 1425, 870.03 and 708 cm^{-1}). The broad band at ~ 1020 - 1016 cm^{-1}arises from C-S-H vibrations, in agreement with those reported by Martinez-Ramirez [52].

As the same for the X-ray diffraction, the FT-IR analysis of composite after attack by different acids (Table 5, Figure 7) confirms the appearance of a trace of calcium chloride ($CaCl_2$) on the specimens exposed to various HCl acids. The low quantity of calcium chloride found is due to its high solubility in water.

Therefore, in Table 5 and Figure 7 no absorption bands corresponding to calcium hydroxide were detected in all of specimens exposed to acidic solutions, which is in agreement with XRD analysis. The $Ca(OH)_2$ was consumed by different HCl acids following the chemical reaction (Equation (3)).

DTA-TGA/DTG Analyses

The simultaneously traced DTA-TGA curves of PET0 composite before and after HCl acids attack (0.5%, 1% and 1.5%) are presented in Figures 8-11.

Table 5: Fourier-transform infrared table of composite before and after attack by different HCl acids, in KBr pellet

Pellet. Positions and Intensities of Infrared Absorption Bands (FT-IR)				
(cm−1)	Group	Compounds	Tap water	H C I (0.5%, 1% and 1.5%)
~3000 - 3700	H2O, OH, hydrogen bonds	Gypsum, CSH	+++	+

~3636 (Fine)	H2O, OH, hydrogen bonds	portlandite	+++	---
3450 - 3485	H2O, OH	AFm, AFt	++	+
1614 - 1621	H2O, v O-H	H2O	+	+
1425 - 1433	v C-O	CaCO3 (Calcite)	+	+
869 - 874	v C-O	CaCO3 (Calcite)	+	+
705 - 710.6	v C-O	CaCO3 (Calcite)	+	+
~1015 - 1029	v Si-O	CSH vibrations	+	+
~470 - 524; 535	v Ca-Cl	CaCl2.2H2O	0	+++
~2970	v C-O (harmonic bands)	CaCO3 (Calcite)	+	+
2865 - 2871	v C-O (harmonic bands)	CaCO3 (Calcite)	+	+
2512 - 2518	v C-O (harmonic bands)	CaCO3 (Calcite)	+	+
~1799	v C-O (harmonic bands)	CaCO3 (Calcite)	+	+

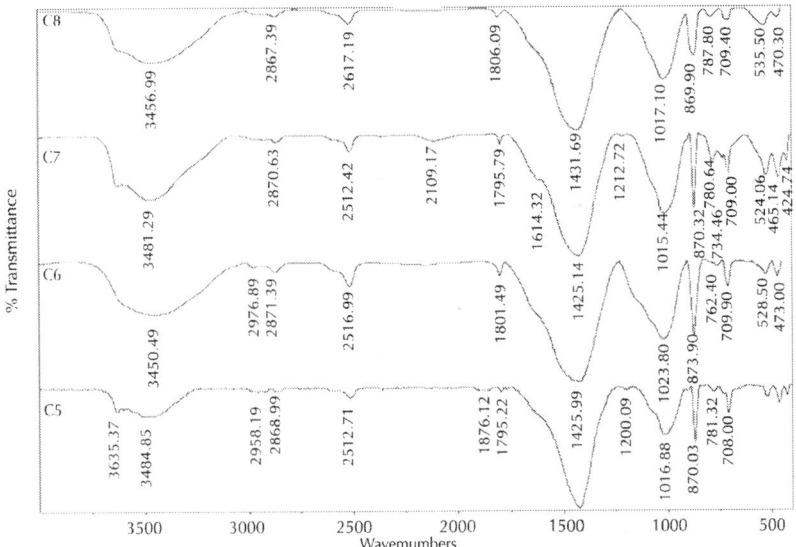

Figure 7: FT-IR spectra of the specimens under HCl acid exposure, (C5: Tap water, C6: 0.5%, C7: 1%, C8: 1.5%).

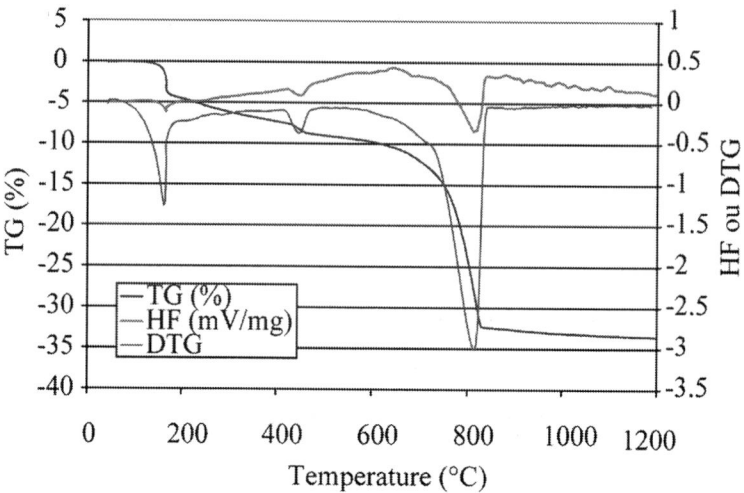

Figure 8: DTA/TG curves at 20 K/min of composite PET0 in tap water exposure.

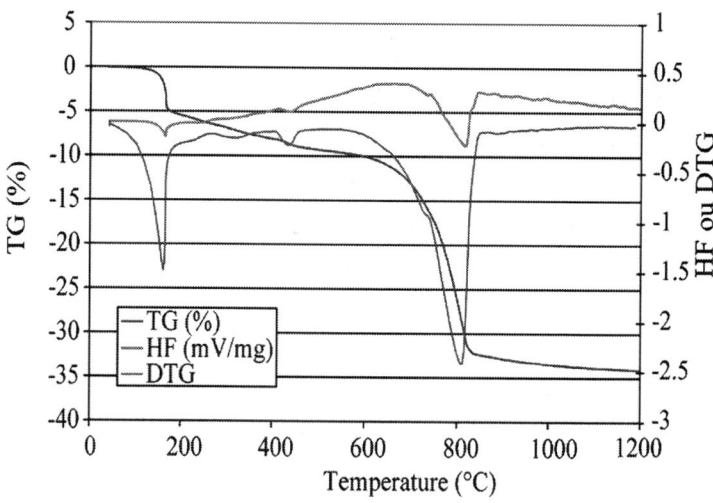

Figure 9: DTA/TG curves at 20 K/min of composite PET0 under 0.5% hydrochloric acid exposure.

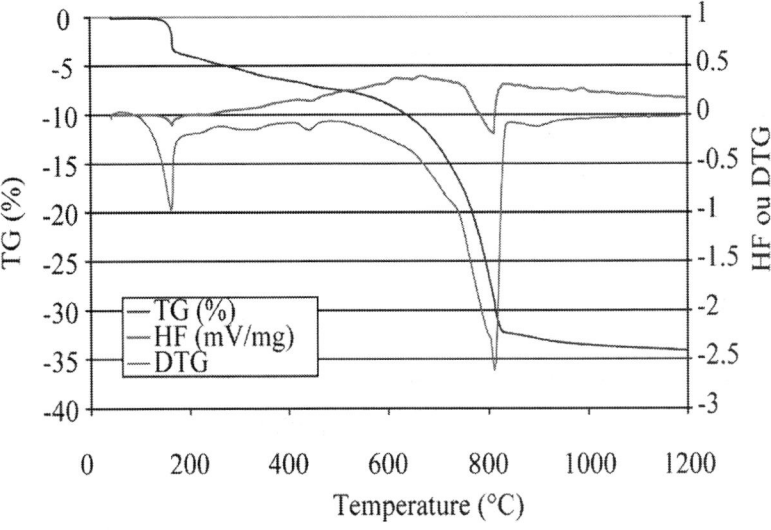

Figure 10: DTA/TG curves at 20 K/min of composite PET0 under 1.0% hydrochloric acid exposure.

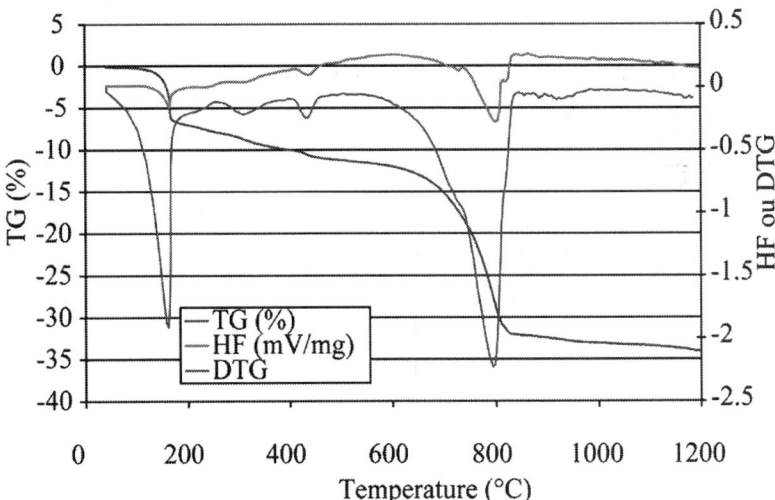

Figure 11: DTA/TG curves at 20 K/min of composite PET0 under 1.5% hydrochloric acid exposure.

Figure 8 shows the DTA/TG curves of mortar without polymer PET0 in the tap water. It can be seen that DTA-TG/dTG curves for this mortar consist of four zones:

~100°C - 150°C: dehydration of pore water (CSH, ettringite)~225°C - 230°C: dehydration of calcium aluminates hydrates~450°C - 550°C: dehydroxylation of calcium hydroxide~700°C - 850°C: decarbonation of $CaCO_3$.

Figures 9-11 show DTA-TG/dTG curves of PET0. As it is shown, the curves can be divided into four major parts, according to different reactions:

~100°C - 145°C: dehydration of pore water~225°C - 230°C: dehydration of calcium aluminates hydrates~440°C - 540°C: the reduction of the intensity of the peak corresponding to dehydroxylation of portlandite, which indicates that, with exposure all the $Ca(OH)_2$ that was formed reacted with the HCl acid solutions (Equation (3)).

~700°C - 850°C: decarbonation of $CaCO_3$.

All the weight loss data are expressed as a function of the ignited weight of the sample, as suggested by Taylor [53]. The calcium hydroxide content was determined from the following equation:

$$CH\left(\%\right) = WL_{CH}\left(\%\right) \times \frac{MW_{CH}}{MW_{H}}$$

(6)

where CH(%) is the content of $Ca(OH)_2$ (in weight basis), WL_{CH}(%)is the weight loss occurred during the dehydration of calcium hydroxide (in weight basis), MW_{CH} is the molar weight of calcium hydroxide and MW_H is the molar weight of water.

When one attacks the PET0 composite by the hydrochloric acid at 0.5%, 1% and 1.5%, there is a diminution of the calcium hydroxide (CH) content, as can be seen in Figure 12. Because when the concentration of acid increases, there is less formation of portlandite $Ca(OH)_2$, which is mainly consumed following the chemical reaction (Equation (3)). This is confirms our previous results.

Scanning Electron Microscope (SEM) Observations

In case of HCl attack, the damaged of PET0 composite was selected as an example to observe the microstructure of deterioration and the SEM images are shown in Figure 13. No portlandite crystal exists in the acid attacked composite. It is clear that exposure to solution of hydrochloric acid generates pores and micro-cracks in the mortar specimens. So, The SEM and ATD-TG/dTG analyses corroborate some of the results discussed above.

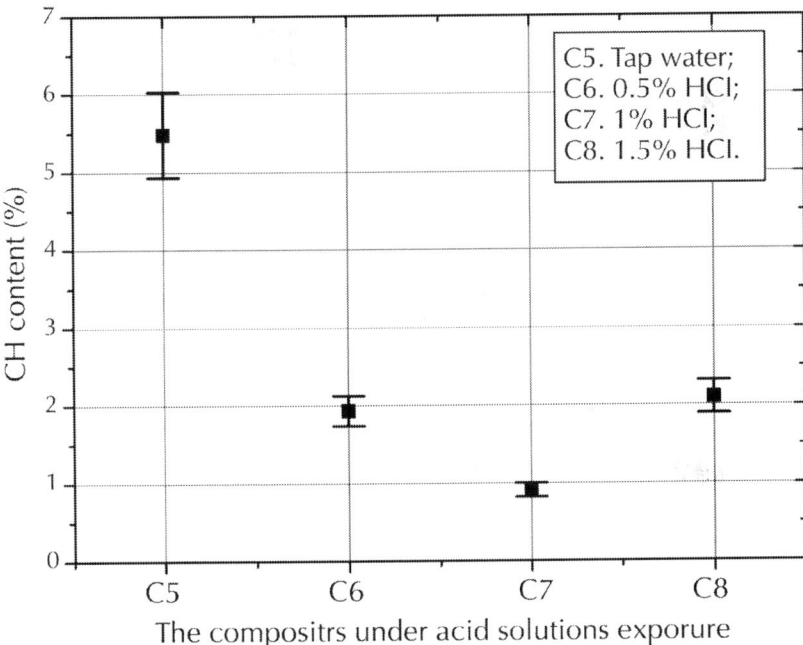

Figure 12: Effect of HCl acids concentration on the calcium hydroxide content (CH).

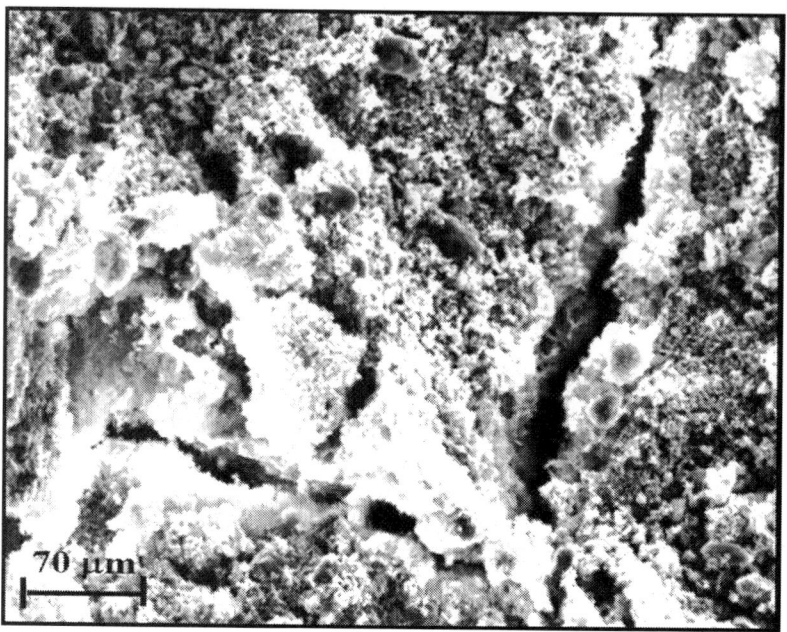

Figure 13: SEM images obtained for the external porous texture of PET0 sample after being attacked by the 1% hydrochloric acid solution (×600).

Ammonium Chloride NH$_4$Cl Attack

Figures 14 and 15 show the evolution of the degraded thickness of composite immersed in a solution of ammonium chloride 10%, represented as a function of the polymer-cement ratios. Loss of mass followed the same trend. At 480 days, significant differences were obtained: compared to PET0, the depth of attack was 16.2%, 26.2% and 34.8% lower for PET2.5, PET5 and PET7.5, respectively.

Ammonium salts are generally more destructive than salts of other bases [54]. So, ammonium chloride is very aggressive and reacts following an exchange mechanism $2NH_4^+ \rightarrow Ca^{2+}$ given by the reaction:

$$Ca(OH)_2 + 2NH_4Cl \rightarrow CaCl_2 + 2H_2O + NH_3\,gas$$

$$\left(NH_4OH = H_2O + NH_3\,gas\right)$$

$$(7)$$

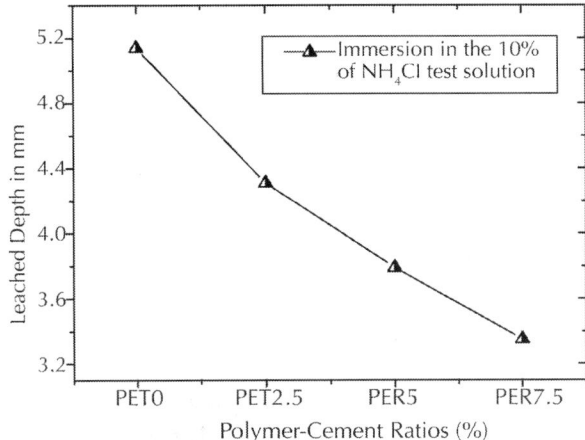

Figure 14: Depth of attack due to the leaching of composite immersed in 10% ammonium chloride.

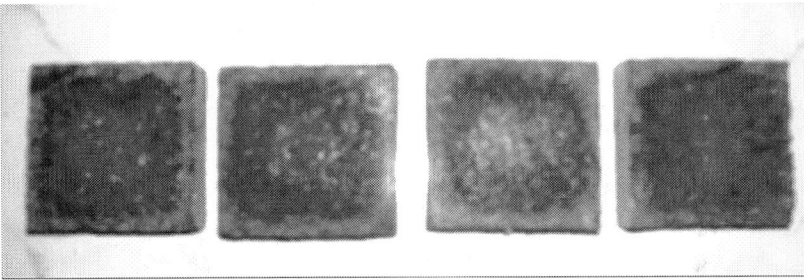

Figure 15: Depth of reduced lixiviation of the composites PET0, PET2.5, PET5 and PET7.5 (from the left to right) indicated by the phenolphthalein solution.

This reaction leads to the formation of a highly soluble calcium chloride ($46.08g/100ml$ of H_2O) and a release of ammonia. The resulting reduction of the pH impedes the reaction to reach a state of equilibrium. The conesquence is progressive leaching of portlandite and C-S-H, thus leading to a decrease of the mechanical properties of mortar. The reaction with aluminates leads to the formation of a hydrated calcium chloride aluminate $3CaO.Al_2O_3.CaCl_2.10H_2O$ [55].

As for most of the other tests, the composites were the modified mortar most resistant to this acid attack (pH = 5.6). This might be due to impervious PET granules blocking the passage of the aggressive solutions and the reduction of the sorptivity of modified mortar (Tables 3 and 4). Furthermore, the decrease in porosity due to the incorporation of PET in composite [48] contributes to reduce the absorption of acidic solution accompanied by a reduction of loss in weight. However, no pertinent data were found concerning the effect of PET against ammonium chloride attack.

Ammonium Sulfate $(NH_4)_2SO_4$ and H_2SO_4 Attacks

Figure 16 shows the percentage change with respect to time in the weight of cubes immersed in sulfuric acid and ammonium-sulfate. It can be seen in the same figure that there is an increase in weight of cubes immersed in ammonium sulfate for the period of experimentation. The augmentation in weight after 120 days of immersion was 4.5%, 4.8%, 4.86% and 6.9% for PET0, PET2.5, PET5 and PET7.5 respectively. However, in the cubes immersed in sulfuric acid it can be noted that the reduction in weight has occurred. So, the rate of reduction in weight was less in PET5 and PET7.5. The mass loss for PET7.5 is lower than the corresponding PET0 mortar by 5% (Figure 16).

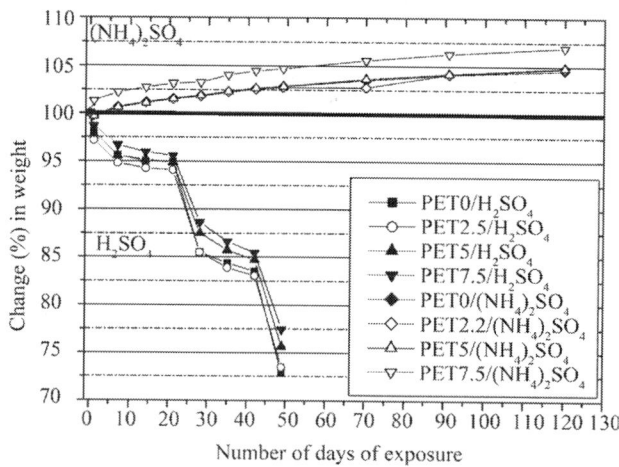

Figure 16: Change (%) in weight with time under H_2SO_4 acid and $(NH_4)_2SO_4$ solutions.

The increase in the weight of composites immersed in sulfatic solutions can be explained by the formation of expansive products. In the presence of water, the sulfates ions react with calcium aluminates hydrate (C_3A) and/or the components of the calcium hydroxide of hardened cement paste to form calcium sulfoaluminate hydrate commonly called "ettringite" [56] and calcium sulfate (gypsum). However, it is important to recognize that the end products of the various reactions, if they occur so as to damage the mortar or concrete, result in different types of damage [57]. Mehta [58] reported that the 5% $(NH_4)_2SO_4$ solution proved to be more aggressive than 1% H_2SO_4; it appears that ammonium salts are able to decompose the calcium hydrate, which is the principal solid phase in hydrated portland cement pastes. In consequence, different preventive measures are required.

Compressive strength losses CSL% of PET-composites samples after curing in water for 28 days and exposed in 10% of ammonium-sulfate solution for 860 days and in 5% of sulfuric acid for 56 days are presented in Figure 17. It can be seen that the modified mortar samples containing polyethylene therephtalate have better behaviour in ammonium-sulfate solution. The results indicate that the resistance to $(NH_4)_2SO_4$ solution of the composites was increased with an increase in PET content. At day-860, the CSL% of PET2.5, PET5 and PET7.5 were reduced by 19.2%, 18.2% and 22.6%, respectively, when compared to that of PET0. In addition, at 56 days, there is a CSL% diminution around 7% for PET7.5 compared to an unmodified one in sulfuric acid solution (Figure 17).

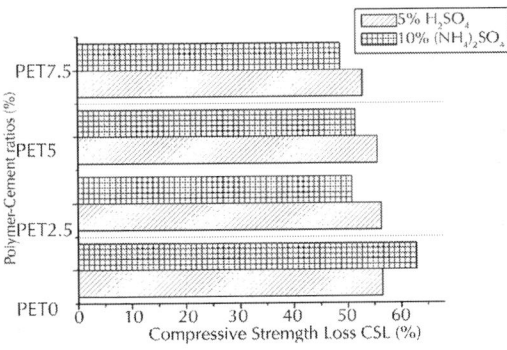

Figure 17: Compressive strength loss of specimens under H_2SO_4 acid and $(NH_4)_2SO_4$ solutions exposure.

Therefore, the increase in the resistance to ammonium-sulfate of the composites is attributed to the impervious PET granules blocking the passage of the aggressive solution and the reduction of the sorptivity of polymer-mortar (Tables 3 and 4). Consequently, the decrease in porosity due to the incorporation of PET in modified mortars [48] contributes to reduce the absorption of sulfates solution. These results are in agreement with those reported elsewhere [46]. The same explanation was also done for the PET7.5 in the sulfuric acid. Additionally, different teams of researchers [49-51] reported that the incorporation of organic additions (polymers) increases chemical resistance in aggressive media.

It can be concluded that adding PET to blended portland cement makes this cement become more resistant to the ammonium-sulfate aggressive environment. It is evident that the resistance of cement to sulfate aggression is also related to the content of C_3A in them [27]. However, no pertinent data were found concerning the effect of PET against ammonium sulfate attack.

Ammonium sulfate is very aggressive and reacts following an exchange mechanism $2NH_4^+ \rightarrow Ca^{2+}$ given by the reaction:

$$Ca(OH)_2 + (NH_4)_2 SO_4$$
$$\rightarrow CaSO_4 + 2H_2O + 2NH_3 gas.$$
$$(NH_4OH = H_2O + NH_3 gas) \tag{8}$$

This reaction leads to the formation of less watersoluble gypsum $CaSO_4$, 0.241 g/100 ml of H_2O and a release of ammonia. Calcium sulfate formed as described above can subsequently react with C_3A, usually via the formation of mono-sulfoaluminate, to form ettringite (Equation (5)). In addition, attack of hydrated Portland cement by H_2SO_4 acid is two-fold. The first one is by acid attack or hydrogen ions and the second is by the sulfuric ions. Two salts are formed: namely calcium sulfate and ettringite (Equations (4) and (5)). These are destructive salts and the pressure produced during their formation causes mortar to crack and disintegrate.

Comparing the exposure of mortars to sulfuric acid and ammonium sulfate, it is immediately apparent that the mechanisms involved are quite different. So, Rendell et al. [59] reported that sulfuric acid causes a heavy deposition of gypsum that acts as a protective layer; this surface protection is verified by the unchanged hardness beyond a depth of 2 mm, however, it is noted that gypsum is deposited at depths of up to 2 mm in fissures and voids. In this case the mechanism of attack is caused by the production of the expansive gypsum, this expansive effect being responsible for the progressive opening of the material structure by dislocation of surface material. It was noted from the SEM and microanalysis [59], that there is a filling of fissures and pores with gypsum, this deposition will have the effect of producing an internal blocking of the pore structure. It is, therefore, proposed that the surface deposit of gypsum, if not removed by scour forms a protective layer thus limiting the attack rate.

The samples exposed to ammonium sulfate experienced a significant deterioration in material hardness up to a depth of 5 mm [59]. However, from the microanalysis it was noted that there was little sulfur evident behind the surface; the major site of gypsum formation being at the surface. It is proposed that the mechanism of deterioration in this case is one of dissolution of Ca^{2+} from the mortar. Leaching of Ca^{2+} is principally the result of the dissolution of portlandite; this action causes an opening of the pore structure of the mortar and is, therefore, responsible for the reduction in mechanical strength and ultimate loss in durability [59].

Visual Inspection

Additionally, in case of H_2SO_4 and $(NH_4)_2SO_4$ attacks, the damaged of PET0 composite was selected as an example to observe the macrostructure of deterioration. So, these results are confirmed by the change of surface samples before and after immersion in the aggressive solutions. A visual inspection of specimens as shown in Figure 18 revealed the deterioration of the samples, particularly for the mortars immersed in sulfuric acid. These mortars kept their cubic forms more or less, but their dimensions decreased considerably.

Photos of specimens stored in the ammonium sulfate solution for 120 and 860 days are presented inFigure 18. The samples stored for 120 days under ammonium sulfate exposure showed the first signs of deterioration, while the specimens stored in tap water did not show any clear evidence of attack. The discussion below concerns the samples stored in $(NH_4)_2SO_4$ solution. In all cases, the first sign of attack was the deterioration of the corners, followed by cracking along the edges. Progressively, expansion and spalling took place on the surface of the specimens.

(a)

(b)

(c)

(d)

Figure 18: Specimens cubes (50 mL) exposed to: (a) Tap water, (b) H_2SO_4 (56 days), (c) $(NH_4)_2SO_4$ (120 days), (d) $(NH_4)_2SO_4$ (860 days).

After 860 days of immersion in a solution of 10% $(NH_4)_2SO_4$, all 4 surfaces of the specimens developed a white cover that was friable and started to peel off, leaving the aggregates uncovered and reducing the connectivity of the paste. Thus, the extent of surface deterioration after 860 days of exposure had a tendency to decrease with the increased replacement level of the PET wastes.

Polymer-mortar composites modified with PET waste can be advantageous for special applications where the main request is not for mechanical properties, such as in the production of sound barriers and cement blocks for lightweight concrete walls. Also, these composites are often used as low-cost materials for preventing chemical attacks or repairing various reinforced concrete structures damaged by chloride-induced corrosion [21,22], as well as in structures exposed to aggressive environments where high resistance to ammonium-sulfate/-chloride, acid, basic and chloride solutions is required.

X-Ray Diffraction (XRD) Analysis

Figure 19 presents the XRD analysis of PET0 composite, as an example, before and after attack by different sulfuric acid and ammonium sulfate solutions. The common factor for all hydrated samples is the presence of portlandite $Ca(OH)_2$ and ettringite $Ca_6[Al(OH)_6]_2(SO_4)_3.26H_2O$ Unlike thesam ples from the sulfuric acid and ammonium sulfate solution which are marked by considerable quantities of gypsum $(CaSO_4.2H_2O)$, the samples which were in water contain minimal quantity of gypsum. As gypsum existed in the initial material in both cases, it is evident that during the process of hydration it served, together with other components, to form ettringite, for water cured samples.

Opposite to this, the quantity of formed ettringite is small and similar in all the samples exposed to ammonium sulfate, which points out to the assumption that mostly sulfate ions only from gypsum present in the initial samples participated in the forming of ettringite. These results are in agreement with those reported by Mileti et al. [27]. We also noted that portlandite $Ca(OH)_2$ was completely decomposed by different acid and sulfate solutions following the chemical reactions 4, 5 and 8.

FT-IR Analyses

Table 6 illustrates the positions and intensities of infrared absorption bands and Figures 20 and 21 show the FT-IR patterns of specimens exposed to: Tap water, H_2SO_4, and $(NH_4)_2SO_4$. The FT-IR spectra of the composite hydrated up to 28 days and cured in water are presented in Figure 20and Table 6. The major changes of the FT-IR spectra in the hydrated cement pastes are: Calcium hydroxide bands ()($-1\sim3637cm^{-1}$) and also for the free OH groups, combined and adsorbed water of CSH, AFm and AFt phases)($3452cm^{-1}$), molecular water (3425-3445 and 1641-$1660cm^{-1}$), carbonate phases .($\sim1428,874.5$ and $712.5cm^{-1}$) The broad band at ~990 - 1020 cm^{-1} arises from C-S-H vibrations, in agreement with those reported by Martinez-Ramirez [52].

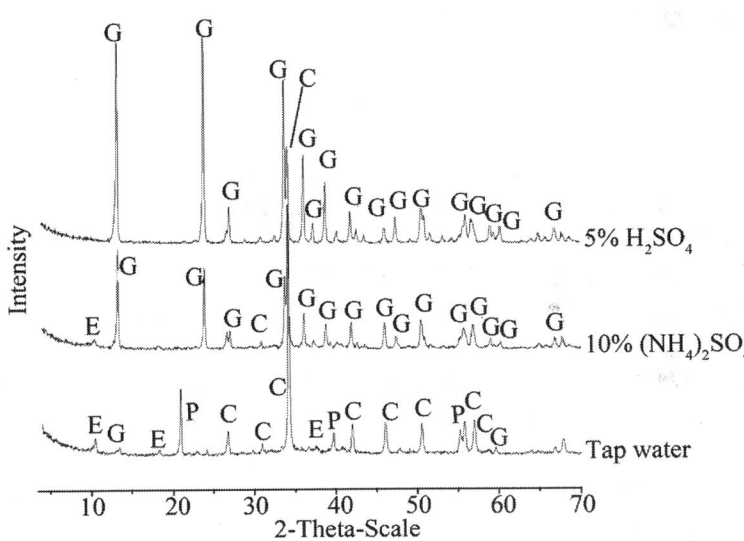

Figure 19: X-ray diffraction pattern of the specimens under H_2SO_4 acid and $(NH_4)_2SO_4$ solutions exposure. P: portlandite $Ca(OH)_2$; E: ettringite; C: calcite; G: gypsum.

As the same for the X-ray diffraction, the FT-IR analysis of composites after attack by acid and sulfate solutions (Table 6, Figures 20 and 21) confirms the appearance of a large quantity of gypsum ($CaSO_4.2H_2O$) on the specimens exposed to H_2SO_4 and $(NH_4)_2SO_4$ and a small quantity of ettringite on the specimens exposed to ammonium sulfate.

Therefore, in Table 6 and Figures 20 and 21 no absorption bands corresponding to calcium hydroxide were detected in all of specimens exposed to acidic and ammonium sulfate solutions, which is in agreement with XRD analysis. The $Ca(OH)_2$ was consumed by aggressive solutions following the chemical reactions 4, 5 and 8.

To obtain further evidence to corroborate these observations from XRD and FT-IR, further investigation was made using SEM, DSC and TG/dTG analyses.

Scanning Electron Microscope (SEM) Observations

The composite of PET0 exposed to $(NH_4)_2SO_4$ solutions was selected as an example to observe the microstructure of deterioration products, and the SEM images are shown in Figure 22. No portlandite $Ca(OH)_2$ crystal existed in the sulfate-attacked unmodified mortar, and many club-shaped or needle-like crystals were embedded irregularly in the pulpy material with a very open microstructure. Figure 22(a) shows a large number of needlelike ettringite crystals, smaller than 2μm in diameter and 10 - 11 μm in length, and the existence of different morphologies of the C-S-H gel in the specimen covering the mortar surface before sulfation. In the pores or surface cracks, there were many club-shaped gypsum crystals, which had much larger sizes (smaller than 100 - 200 μm in diameter and 300 - 500 μm in length) than the ettringite, as shown in Figure 22(b). The formation of both ettringite and gypsum led to the cracking, spallingand decomposition of the unmodified mortar exposed to a sulfate environment (Figure 22(c)). The combination of XRD, FT-IR and SEM led to the positive identification of the deterioration products' formation.

The surface that was exposed to sulfuric acid had a nature that was quite different. The scanning electron microscopy (SEM) observations were also carried out to identify the products formed by attack of PET0 in the sulfuric acid solutions and the results are exhibited in Figure 23. In addition, Figure 23 taken around a crack on the surface of the core sample shows deposition of a large number of needle-like crystals (gypsum), smaller than 36 - 40 μm in diameter and 300 - 500 μm in length, covering the mortar surface. In the pores or surface cracks, there are many club-shaped gypsum crystals as shown in Figure 23. The formation of gypsum leads to the cracking, spalling, and decomposition of mortar exposed to sulfuric acid environment. The SEM images obtained by SEM corroborate some of the results discussed above.

Table 6: Fourier-transform infrared table of composite before and after attack by different acid and sulfate solutions, in KBr pellet

Positions and Intensities of Infrared Absorption Bands (FT-IR)					
(cm−1)	Group	Compounds	Tap water	H2SO4	(NH4)2SO4
~3000 - 3600	H2O, OH, hydrogen bonds	Gypsum, CSH	++	+	+
~3637.09	H2O, OH, hydrogen bonds	portlandite	++	0	0
3424 - 3452	H2O, OH	AFm, AFt	+	+	++
1623 - 1660	H2O, OH	H2O	+	+	+
1428 - 1439	v C-O	CaCO3 (Calcite)	+	+	+
873 - 875	v C-O	CaCO3 (Calcite)	+	+	+
710 - 713	v C-O	CaCO3 (Calcite)	+	+	+
~990 - 1020	v Si-O	CSH vibrations	+	+ (w,sh)	+
~1130 - 1142	v S-O	Gypsum	~	+++	+++
~604.5 - 672	S-O	Gypsum	~	+++	+++
~3408 - 3546	δ O-H	water (Gypsum)	~	+++	+++
~1622 - 1687	δ O-H	water (Gypsum)	~	+++	+++
~1105 - 1110	δ S-O (sulfates)	Ettringite	+	0	~
~602 - 608	δ S-O (sulfates)	Ettringite	+	0	~
2982	δ C-O (harmonic bands)	CaCO3 (Calcite)	+	+	+
2872	δ C-O (harmonic bands)	CaCO3 (Calcite)	+	+	+
~2515	δ C-O (harmonic bands)	CaCO3 (Calcite)	+	+	+
~1798	δ C-O (harmonic bands)	CaCO3 (Calcite)	+	+	+

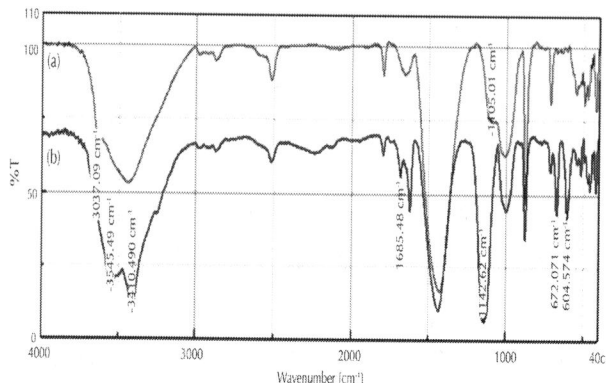

Figure 20: FT-IR spectra of the specimens under (a) Tap water and (b) $(NH_4)_2SO_4$ solution exposure.

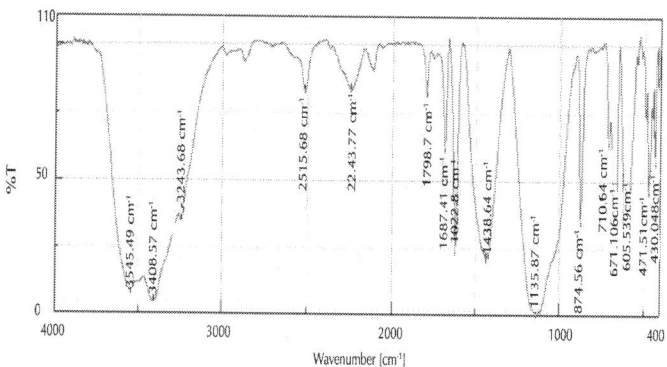

Figure 21: FT-IR spectra of the specimens under sulfuric acid exposure.

Consequently, our results are in agreement with those reported by Rendell et al. [59] which showed that when cement mortar is exposed to sulfuric acid a dense layer of gypsum is formed; this is capable of retarding the deterioration process by acting as a surface sealing layer. Gypsum also exists in pores and fissures in the surface zone, indicating that the attack is due to expansive crystallisation. In ammonium sulfate the surface deposit of gypsum is sparse and the damage to the concrete occurs to a greater depth. The lack of sulfur found in the surface zone indicates that the mechanism of deterioration is due to the dissolution of Ca^{2+}.

DSC and TG/dTG Analyses

The DSC and TG/dTG curves of PET0 composite before and after $(NH_4)_2SO_4$ solution and H_2SO_4 acid attack are presented in Figures 24-26.

Figure 24 shows the DSC curves of unmodified mortar PET0 in the tap water. It can be seen that DSC curves for this mortar consist of three zones:

~90°C - 115°C: dehydration of pore water (CSH, ettringite)~145°C - 157°C: dehydration of calcium sulphate (Gypsum)~450°C - 500°C: dehydroxylation of calcium hydroxide, portlandite.

Figure 25 presents the DSC trace of PET0 after ammonium sulphate attack; it displays four endothermic peaks, according to different reactions:

~100°C - 145°C: dehydration of pore water~115°C - 120°C: ettringite $Ca_6[Al(OH)_6]_2(SO_4)_3.26H_2O$ ~140°C - 145°C: the intensity of the peak due to calcium sulphate $CaSO_4.2H_2O$ is greater. So, gypsum was the dominant reaction product (Equation (8)) while the ettringite appeared as a trace element in the composite.

~440°C - 510°C: In PET0 composite there was no endothermic peak at 462°C, which indicates that, with exposure all the $Ca(OH)_2$ that was formed reacted with the ammonium sulphate solution.

Figure 26 presents the TG/dTG trace for the surface part of the sample obtained from the PET0 composite stored in 5% sulphuric acid solution; it displays three endothermic peaks at 157 (High intensity), 210 and 850°C - 900°C, indicating gypsum, calcium monocarpboaluminate hydrated and decarbonation of the calcite $CaCO_3$. Also, TG/dTG analysis showed no endothermic peak at 462°C, which indicates that, with exposure all the $Ca(OH)_2$ that was formed reacted with the sulphuric acid solution (Equation (4)).

Hence, the DSC, TG/dTG curves obtained by DSC and TG/dTG analyses corroborate some of the results discussed above.

It should be pointed out that, in practice, the durability of composites exposed to the chemical solutions investigated in this study should be much better than indicated by the test data given here. First, because the chemical solutions of high concentrations used in the test are not commonly encountered in food and most other industries. Second, the composite specimens in the test were fully submerged, whereas in the industrial practice the structural element such as floor is usually exposed to attack from one side only. Third, the immersion solutions were kept in state of constant motion, and were frequently replaced which fresh brushing of the specimens once every week; this removed the reaction products and almost continuously forced new material to come into contact with aggressive solution.

2012/11/26 13:30 L D6.9 ×5.0k   20 um

(a)

2012/11/26 13:26 L D6.1   ×180   500 um

(b)

2012/11/26 13:43 1. D6.8 ×150   500 um

(c)

Figure 22: SEM photos of the PET0 composite after exposure to 10% $(NH_4)_2SO_4$ solution for 860 days; (a) Needlelike ettringite crystals, (b) Club-shaped gypsum crystals and (c) deteriorate mortar.

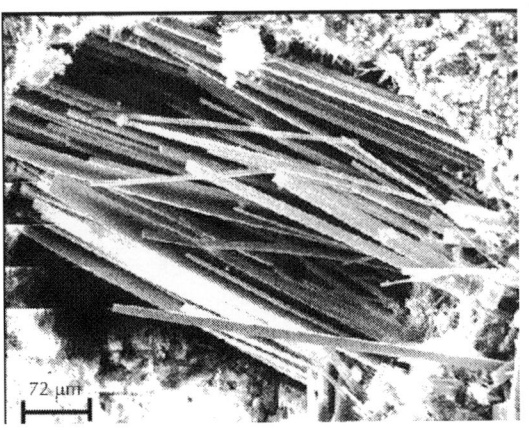

72 µm

Figure 23: SEM photos of the PET0 composite after exposure to 5% H_2SO_4 solution (×600) for 56 days; needle-like crystals (gypsum) in the pore.

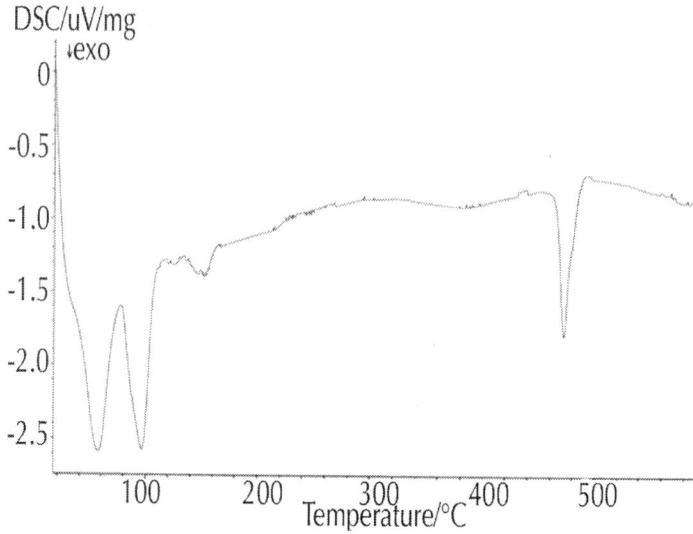

Figure 24: DSC curves at 10 K/min of composite PET0 in tap water exposure.

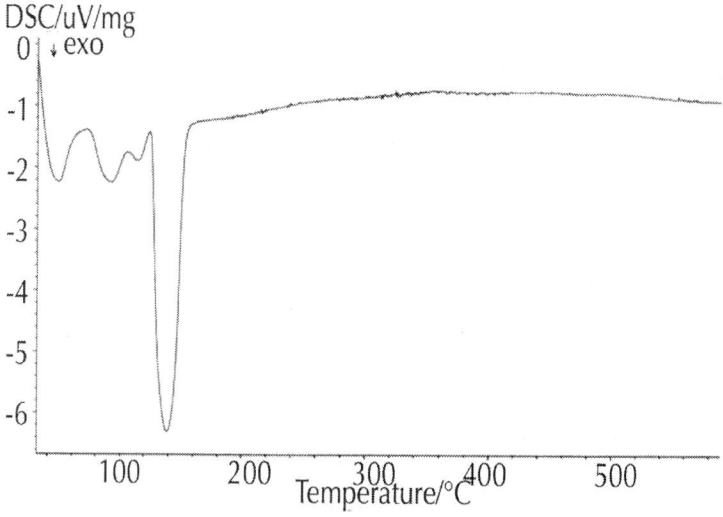

Figure 25: DSC curves at 10 K/min of composite PET0 in $(NH_4)_2SO_4$ solution exposure.

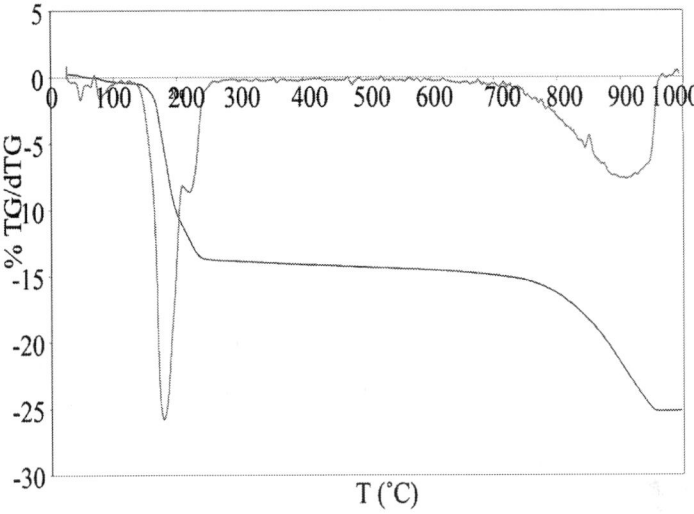

Figure 26: TG/dTG curves at 10 K/min of composite PET0 in H_2SO_4 acid exposure.

CONCLUSIONS

The influence of the PET particles on PET-modified mortar composites sorptivity, mechanical strength and the chemical properties under hydrochloric acid, ammonium-chloride, ammonium-sulfate and sulfuric acid attack solutions were determined. The results of testing the attack by aggressive solutions allow the following conclusions:

- Test results of the hydraulic transport properties revealed that the addition of PET particles tends to restrict water propagation in the cement matrix and reduces water absorption of the composite. The decrease of the sorptivity-value is favorable to the durability of the specimen structures.

- The results of specific indicators for ammonium chloride, ammonium-sulfate and acid attacks showed that an increase in the PET content led to better resistance of the composite. This may to the reduced volume of largesized pores and the improved resistance to the absorption of aggressive solutions with the addition of PET in the cement pastes, because of the impervious

PET particles blocking the passage of the aggressive ions. In addition, to the reduction in the porosity near particle/matrix interfacial zone, due to the high bonding between PET additive and cement paste.

- Among composite mixes in the HCl acid attacks, PET7.5 mixes perform better than other mixes. The chemical resistance characteristics of materials are affected by the concentration and by the nature of acids following the order:

  $0.5\%HCl < 1.0\%HCl < 1.5\%HCl$ Additionally, from the simultaneously traced DTA/TG curves there is a diminution of the calcium hydroxide (CH) content, because of the formation of calcium chloride which has a high solubility in water.

- On the basis of the XRD, FT-IR and DSC analyses of the changes in the phase composition, it can be concluded that the main product of the ammonium-sulfate corrosion of cement is gypsum, accompanied by ettringite. The common factor for all hydrated samples is the presence of portlandite $Ca(OH)_2$ and ettringite

  $Ca_6Al_2(SO_4)_3(OH)_{12} \cdot 26H_2O$. The samples exposed to the ammonium-sulfate solution show considerable quantities of gypsum $CaSO_4 \cdot 2H_2O$. The quantity of ettringite formed is small and very similar in all the samples, which points out to the assumption that mostly sulfate ions only from gypsum present in the initial samples participated.

- The SEM results show that the presence of sulfur ions is virtually existent in the matrix of composites exposed to ammonium sulfate. It is proposed that the reduction in mechanical hardness can be attributed to a migration of Ca^{2+} from the outer layers of the unmodified mortar.

- The mechanism of attack in the case of composite exposed to sulfuric acid is principally a surface phenomenon, the expansive action of the gypsum acting to dislocate aggregate and to increase micro cracking at the surface. It is evident from the SEM that there is a filling of pores and micro fissures with gypsum (Tg/dTG). In the absence of erosion this layer has a surface blocking ability.

The utilization of the PET waste particles as a binder instead of cement in the manufacture of PET-mortar composites and as a sustainable building materials in the construction industry help to preserve natural resources and maintain the ecological balance and also to prevent or repair various reinforced concrete structures.

ACKNOWLEDGEMENTS

The authors acknowledge the financial support from the Ministry of Higher Education and Scientific Research of Algeria, under the grants CNEPRU J0405520120005. The authors greatly appreciated the technical support from the Laboratory of LUSAC EA 2607, University of Caen Basse-Normandie, Cherbourg Octeville (French) and also would like to thank Mr M. T. Gouasmi and the graduate students, Mrs. A. Herizi and A. Kerour for their help.

REFERENCES

1. S. Mileti , M. Ili , J. Ranogajec, R. Marinovi -Neducin, and M. Djuri , "Portland Ash Cement Degradation in Ammonium-Sulfate Solution," Cement and Concrete Research, Vol. 28, No. 5, 1998, pp. 713-725. doi:10.1016/S0008-8846(98)00023-4

2. Y. Ohama, "Polymer Based Admixtures," Cement and Concrete Composite, Vol. 20, No. 2-3, 1998, pp. 189-212. doi:10.1016/S0958-9465(97)00065-6

3. Y. Ohama, "Hand Book of Polymer-Modified Concrete and Mortars, Properties and Process Technology," Noyes Publications, Park Ridge, 1995, p. 236.

4. D. W. Fowler, "Polymers in Concrete: A Vision for the 21st Century," Cement and Concrete Composites, Vol. 21, No. 5-6, 1999, pp. 449-452. doi:10.1016/S0958-9465(99)00032-3

5. N. Saikia and J. de Brito, "Use of Plastic Waste as Aggregate in Cement Mortar and Concrete Preparation: A review," Construction and Building Materials, Vol. 34, 2012, pp. 385-401. doi:10.1016/j.conbuildmat.2012.02.066

6. R. Siddique, J. Khatib and I. Kaur, "Use of Recycled Plastic in Concrete: A review," Waste Management, Vol. 28, No. 10, 2008, pp. 1835-1852.doi:10.1016/j.wasman.2007.09.011

7.   K. F. Portella, A. Joukoski, R. Franck and R. Derksen, "Reciclagem Secundária de Rejeitos de Porcelanas Elétricas em Estruturas de Concreto: Determinação do Desempenho sob Envelhecimento Acelerado," Cerâmica, Vol. 52, No. 323, 2006, pp. 155-167. doi:10.1590/S0366-69132006000300008

8.   I. Guerra, I. Vivar, B. Llamas, A. Juan and J. Moran, "Eco-efficient Concretes: The Effects of Using Recycled Ceramic Material from Sanitary Installations on The Mechanical Properties of Concrete," Waste Management, Vol. 29, No. 2, 2009, pp. 643-646.doi:10.1016/j.wasman.2008.06.018

9.   Z. Z. Ismail and E. A. Al-Hashmi, "Use of Waste Plastic in Concrete Mixture as Aggregate Replacement," Waste Management, Vol. 28, No. 11, 2008, pp. 2041-2047.

10.   S. C. Angulo, C. Ulsen, V. M. John, H. Kahn and M. A. Cincotto, "Chemical-Mineralogical Characterization of C&D Waste Recycled Aggregates from São Paulo, Brazil," Waste Management, Vol. 29, No. 2, 2009, pp. 721-730.

11.   C. Hoppen, K. F. Portella, A. Joukoski, E. M. Trindade and C. V. Andreóli, "Uso de Lodo de Estação de Tratamento de Agua Centrifugado, em Matriz de Concreto de Cimento Portland para Reduzir o Impacto Ambiental," Química Nova, Vol. 29, No. 1, 2006, pp. 79-84.

12.   K. S. Rebeiz, "Time-Temperature Properties of Polymer Concrete Using Recycled PET," Cement and Concrete Composites, Vol. 17, No. 2, 1995, pp. 119-124. doi:10.1016/0958-9465(94)00004-I

13.   Y.W. Choi, DJ. Moon, YJ. Kim, M. Lachemi, "Characteristics of Mortar and Concrete Containing Fine Aggregate Manufactured from Recycled Waste Polyethylene Terephthalate Bottles," Construction and Building Materials, Vol. 23, No. 8, pp. 2829-2835.doi:10.1016/j.conbuildmat.2009.02.036

14.   Y. W. Choi, D. J. Moon, J. S. Chung and S. K. Cho, "Effects of PET Waste Bottles Aggregate on the Properties of Concrete," Cement and Concrete Research, Vol. 35, No. 4, 2005, pp. 776-781. doi:10.1016/j.cemconres.2004.05.014

15.   B. JO, G. Tae and C. Kim, "Uniaxial Creep Behavior and Prediction of Recycled-PET Polymer Concrete," Construction Building Materials, Vol. 21, 2007, pp. 1552-1559.doi:10.1016/j. conbuildmat.2005.10.003

16. T. Ochi, S. Okubo and K. Fukui, "Development of Recycled PET Fiber and Its Application as Concrete Reinforcing Fiber," Cement and Concrete Composites, Vol. 29, No. 6, 2007, pp. 448-455. doi:10.1016/j.cemconcomp.2007.02.002

17. A. Khaloo, M. Dehestani and P. Rahmatabadi, "Mechanical Properties of Concrete Containing a High Volume of Tire-Rubber Particles," Waste Management, Vol. 28, No. 12, 2008, pp. 2472-2482. doi:10.1016/j.wasman.2008.01.015

18. J. C. A. Galvão, K. F. Portella, A. Joukoski, R. Mendes and E. S. Ferreira, "Use of waste polymers in concrete for repair of dam hydraulic surfaces," Construction and Building Materials, Vol. 25, No. 2, 2011, pp. 1049-1055. doi:10.1016/j.conbuildmat.2010.06.073

19. R. Wang and C. Meyer, "Performance of Cement Mortar Made With Recycled High Impact Polystyrene," Cement & Concrete Composites, Vol. 34, No. 9, 2012, pp. 975-981.doi:10.1016/j.cemconcomp.2012.06.014

20. V. Corinaldesi, A. Mazzoli, G. Moriconi, "Mechanical Behaviour and Thermal Conductivity of Mortars Containing Waste Rubber Particles," Materials and Design, Vol. 32, No. 3, 2011, pp. 1646-1650. doi:10.1016/j.matdes.2010.10.013

21. A. S. Benosman, H. Taibi, M. Mouli, M. Belbachir and Y. Senhadji, "Diffusion of Chloride Ions in Polymer-Mortar Composites (PET)," Journal of Applied Polymer Science, Vol. 110, No. 3, 2008, pp. 1600-1605. doi:10.1002/app.28587

22. A. S. Benosman, H. Taïbi, M. Belbachir, I. Bahlouli, M. Mouli, Y. Senhadji and D. Houivet, "Resistance of Polymer (PET)-Mortar Composites to Chloride Penetration," Proceedings of 7th Asian Symposium on Polymers in Concrete ASPIC 2012, Istanbul, 3-5 October 2012, pp. 387-395.

23. M. T. Gouasmi, A. S. Benosman, H. Taibi, M. Belbachir and Y. Senhadji, "Elaboration and Characterization of Polymer-Siliceous Sand Composites," 2nd French Meeting on Insulating Materials RFMI-2, Oran, 17-19 December 2012.

24. M. T. Gouasmi, "Effects of Polyethylene Terephthalate Lightweight Aggregates on the Properties of Mortar," Magister Thesis, University of Oran, Oran, 2013, p. 153.

25. "ECO PET," 2007. http://www.ecopet.eu/Domino_english/ecopet.htm

26. The Korea Institute of Resources Recycling, "The Korean Institute of Resources Recycling, Recycling Handbook," The Korea Institute of Resources Recycling, Seoul, 1999.

27. S. Mileti, M. Ili, S. Otovi, R. Foli and Y. Ivanov, "Phase Composition Changes Due to Ammonium-Sulphate: Attack on Portland and Portland Fly Ash Cements," Construction and Building Materials, Vol. 13, No. 3, 1999, pp. 117-127. doi:10.1016/S0950-0618(99)00017-3

28. H. F. W. Taylor, "Crystal Structure of Some Double Hydroxide Minerals," Mineraogicall Magazine, Vol. 39, No. 304, 1973, pp. 247-256. doi:10.1180/minmag.1973.039.304.01

29. S. Mileti, M. Ili, J. Ranogajec and M. Djuri, "Sulphate Corrosion of Portland Cement and Portland Cement Mixed With Fly Ash and Slag as a Function of its Composition," Proceedings of XVI Symposuim on Nordic Concrete Research, Helsinki, 1996, pp. 339-340.

30. S. Mileti and M. Ili, "Sulphate Corrosion of Portland Cement with Various Mineral Compositions, Proceedings of the 13th International Corrosion Congress, Melbourne, 1996, pp. 1-7.

31. V. Zivica and A. Bajza, "Acidic Attack of Cement Based Materials—A Review. Part 1. Principle of Acidic Attack," Construction and Building Materials, Vol. 15, No. 8, 2001, pp. 331-340. doi:10.1016/S0950-0618(01)00012-5

32. P. K. Mehta, "Concrete, Properties and Materials," Prentice-Hall, Upper Saddle River, 1986.

33. W. H. Gutt and W. H. Harrison, "Chemical Resistance of Concrete," Concrete, Vol. 11, No. 5, 1997, pp. 35-37.

34. V. Zivica and A. Bajza, "Acidic Attack of Cement-Based Materials—A Review Part 2. Factors of Rate of Acidic Attack and Protective Measures," Construction and Building Materials, Vol. 16, 2002, pp. 215-222. doi:10.1016/S0950-0618(02)00011-9

35. A. S. Benosman, H. Taïbi, M. Belbachir, I. Bahlouli, M. Mouli, Y. Senhadji and D. Houivet, "Mineralogical Study of Polymer-Mortar Composites with PET Polymer by Means of Spectroscopic Analyses," Materials Sciences and Applications, Vol. 3, No. 3, 2012, pp. 139-150. doi:10.4236/msa.2012.33022

36.   EN 196-3, "Methods of Testing Cement—Part 3: Determination of Setting Time and Soundness," Comité Européen de Normalisation, Brussels, 1995.

37.   EN 196-1, "Methods of testing cement—Part 1: Determination of Strength," Comité Européen de Normalisation, Brussels, 1995.

38.   ASTM C 267-97, "Standard Test Methods for Chemical Resistance of Mortars, Grouts, and Monolithic Surfacing and Polymer Concretes," American Society for Testing and Materials (ASTM) International, West Conshohocken, 1997.

39.   ASTM C1012-04, "Standard Test Method for Length Change of Hydraulic-Cement Mortars Exposed to a Sulfate Solution," American Society for Testing and Materials (ASTM) International, West Conshohocken, 2004.

40.   C. Carde, G. Escadeillas and R. François, "Use of Ammonium Nitrate Solution to Simulate and Accelerate the Leaching of Cement Pastes due To Deionized Water," Magazine Concrete Research, Vol. 181, No. 49, 1997, pp. 295-301.doi:10.1680/macr.1997.49.181.295

41.   N. Kaid, M. Cyr, S. Julien and H. Khelafi, "Durability of Concrete Containing a Natural Pozzolan as Defined by a Performance-Based Approach," Construction and Building Materials, Vol. 23, No. 12, 2009, pp. 3457-3467.doi:10.1016/j.conbuildmat.2009.08.002

42.   Z. T. Chang, X. J. Song, R. Munn, M. Marosszeky, "Using Limestone Aggregates and Different Cements for Enhancing Resistance of Concrete to Sulphuric Acid Attack," Cement and Concrete Research, Vol. 35, No. 8, 2005, pp. 1486-1494.doi:10.1016/j.cemconres.2005.03.006

43.   S. Goyal, M. Kumar, D. S. Sidhu and B. Bhattacharjee, "Resistance of Mineral Admixture Concrete to Acid Attack," Journal of Advanced Concrete Technology, Vol. 7, No. 2, 2009, pp. 273-283. doi:10.3151/jact.7.273

44.   H. Siad, H. A. Mesbah, H. Khelafi, S. Kamali-Bernard and M. Mouli, "Effect Of Mineral Admixture on Resistance to Sulphuric and Hydrochloric Acid Attacks in Self-Compacting Concrete," Canadian Journal of Civil Engineering, Vol. 37, No. 3, 2010, pp. 441-449.doi:10.1139/L09-157

45.   D. Achoura, Ch. Lanos, R. Jauberthie and B. Redjel, "Influence d'une Substitution Partielle du Ciment par du Laitier de Hauts Fourneaux Sur la Résistance des Mortiers en Milieu Acide," Journal de Physique IV France, EDP Sciences, Vol. 118, No. 1, 2004, pp. 159-164.doi:10.1051/jp4:2004118019

46.   A. S. Benosman, M. Mouli, H. Taibi, M. Belbachir and Y. Senhadji, "Resistance of Polymer (PET)-Mortar Composites to Aggressive Solutions," International Journal of Engineering Research in Africa, Vol. 5, No. 1, 2011, pp. 1-15.doi:10.4028/www.scientific.net/JERA.5.1

47.   A. Allahverdi and F. Škvára, "Acidic Corrosion of Hydrated Cement Based Materials—Part 1. Mechanism of the Phenomenon," Ceramics-Silikáty, Vol. 44, No. 3, 2000, pp. 114-120.

48.   A. S. Benosman, "Mechanical Performance and Durability of Cementitious Materials Modified by Adding Polymer (PET)," Ph.D. Thesis, University of Oran, Algeria, 2010.

49.   J.-A. Rossignolo, M.-V.C. Agnesini, "Durability of Polymer-Modified Lightweight Aggregate Concrete," Cement and Concrete Composites, Vol. 26, No. 4, 2004, pp. 375-380.doi:10.1016/S0958-9465(03)00022-2

50.   J. Monteny, N. De Belie, E. Vincke, W. Verstraete and L. Taerwe, "Chemical and Microbiological Tests to Simulate Sulfuric Acid Corrosion of Polymer-Modified Concrete," Cement and Concrete Research, Vol. 31, No. 9, 2001, pp. 1359-1365.

51.   S. Chandra and L. Berntsson, "Lightweight Aggregate Concrete-Science, Technology, and Applications," William Andrew Publishing/Noyes, Chapter 8, 2002, pp. 231-240.

52.   S. Martínez-Ramírez, "Influence of SO_2 Deposition on Cement Mortar Hydration," Cement and Concrete Research, Vol. 29, No. 1, 1999, pp. 107-111. doi:10.1016/S0008-8846(98)00183-5

53.   H. F. W. Taylor, "Studies on the Chemistry and Microstructures of Cement Pastes," Proceedings of the British Ceramic Society, Vol. 35, 1984, pp. 65-82.

54.   C. L. Page and M. M. Page, "Durability of Concrete and Cement Composites," Woodhead Publishing, Cambidge England, 2007, p. 404.

55.   V. Ukraincik, D. Bjecovic and A. Djurekovic, "Concrete corrosion in a nitrogen fertilizer plant," In: P. J. Seredaand and G. G. Litvan,

Eds., Durability of building materials and components, ASTM, Philadelphia, 1978, pp. 397-409.

56. A. Vichot and J.-P. Ollivier, "La durabilité des bétons," Presses de l'École nationale des Ponts et chaussées, ENPC, France, 2008.

57. A. M. Neville, "The Confused World of Sulfate Attack on Concrete," Cement and Concrete Research, Vol. 34, No. 8, 2004, pp. 1275-1296. doi:10.1016/j.cemconres.2004.04.004

58. P. K. Mehta, "Studies on Chemical resistance of low water/cement ratio concretes," Cement and Concrete Research, Vol. 15, No. 6, 1985, pp. 969-978. doi:10.1016/0008-8846(85)90087-0

59. F. Rendell and R. Jauberthie, "The Deterioration of Mortar in Sulphate Environments," Construction and Building Materials, Vol. 13, No. 6, 1999, pp. 321-327.doi:10.1016/S0950-0618(99)00031-8

Synthesis and Characterization of GaN Rods Prepared by Ammono-Chemical Vapor Deposition

Gregorio Guadalupe Carbajal Arízaga[1], Karina Viridiana Chávez Hernández[2], Nicolás Cayetano Castro[3], Manuel Herrera Zaldivar[2], Rafael García Gutiérrez[4], and Oscar Edel Contreras López[2]

[1]Departamento de Química, Universidad de Guadalajara, Guadalajara, México

[2]Centro de Nanociencias y Nanotecnología, Universidad Nacional Autónoma de México, Ensenada, México

[3]Instituto Potosino de Investigación Científica y Tecnológica (IPICYT), División de Materiales Avanzados (LINAN), San Luis Potosí, México

[4]Centro de Investigación en Física, Universidad de Sonora, Hermosillo, México

ABSTRACT

GaN rods were deposited by chemical vapor deposition (CVD) onto sapphire (0 0 0 1) and amorphous quartz. The reactive Ga species in vapor the phase was formed with NH_4Cl and gallium. The unidirectional growth was catalyzed with gold nanoparticles formed onto the substrate prior to the CVD reaction in order to induce a vapor-liquid-solid (VLS) mechanism. However, this method of synthesis seems to be influenced by other growth mechanisms which formed additional depositions of GaN with different morphology than the rods catalyzed by gold nanoparticles. The moieties of GaN that grew in the absence of gold formed branches in the rods or increased the lateral growth of rods resulting in larger diameters than the size of the gold particle that guided the growth.

INTRODUCTION

Gallium nitride is an attractive semiconductor since its 3.4 eV gap band is suitable for use in optical devices and integrated circuits that operate in wavelengths of the blue-violet and ultra-violet regions [1]. Chemical vapor deposition (CVD) is one of the most investigated methods of GaN synthesis involving different types of gallium-containing reagents that form reactive gallium species in the vapor phase. However, this technique has countless factors that must be controlled such as the type of precursors, concentrations, substrate composition, deposition temperature, and the nature of the catalysts to produce a desired morphology. For instance, in the simplest CVD setup to prepare GaN from only metallic gallium and ammonia, the changes in reagent concentration lead to formation of amorphous GaN, pellets or rods [2].

By increasing the number of reagents, the system becomes more complex, although the use of more reagents may be justified if the reactivity of gallium precursors increases, thus affording a lower reaction temperature or if the reagents are less toxic. This is the case of a method to produce bulk GaN under high pressures with an ammonium halide salt (NH_4X, where X = Cl, Br, or I) that sublimes and forms a gallium complex with metallic gallium, and then the complex reacts with NH_3 to form GaN [3]. These reagents, which are inexpensive and easy to handle, have been used in a CVD system resulting in formation of GaN-containing columns [4]. However, the use of NH_4Cl leads to a series of additional reactions that increase lateral growth with only a change of the reaction temperature [5].

On the other hand, a key point to grow GaN rods, i.e., to favor a larger L/D aspect ratio (length to diameter) to that of columns, is a metal catalyst, or foreign element catalytic agent (FECA) [6], which is used as a metal cluster or nanoparticle. The growth mechanism of GaN wires with metal catalysts has been described by Lieber [7,8] and is based on the capacity of the metal to dissolve gallium and nitrogen under the reaction conditions, as would happen with metallic iron or nickel at 900°C [8] or 650°C [9], respectively. This reaction is called the vapor-liquid-solid (VLS) mechanism [10], since gallium and nitrogen are transferred from the vapor (V) phase to the liquid (L) metal cluster phase and then solidify (S) as GaN, retaining the catalytic droplet on the tips of the rods as the special feature of this process.

In this particular metal cluster catalysis, if one element which is to form the rod is not soluble in the metal catalyst, as in the similar case of nitrogen in gold [8, 11] for example, the nanorods are not formed, as confirmed earlier by Duan et al. [7]. Nonetheless, recent experiments have demonstrated the formation of GaN nanowires employing gold clusters as catalyst when triethyl gallium and ammonia were used as the reagents [8, 11].

Our objective in this investigation is to prepare gold clusters in situ through the annealing of a thin film deposited on sapphire (0 0 0 1) substrate and amorphous quartz and determine whether this substrate influences the formation of clusters. Then, we intend to characterize the rods grown with this practical method in order to know their properties and propose suitable applications.

MATERIALS AND METHODS

Chemical Vapor Deposition (CVD)

Reagents used in this work were metallic gallium (Sigma Aldrich, USA, 99.9995%), ammonium chloride (Fagalab, Mexico, 99.5%), and ammonia gas (Praxair, México, 99.99%). Sapphire (0 0 0 1) and amorphous quartz substrates areas of 1 - 1.5 cm^2 were cleaned with acetone in an ultrasound bath, dried in air, and then transferred to a vacuum chamber (JEOL-JEE-400) where a gold film of ~10 nm in thickness was deposited by vacuum sputtering onto the surfaces of the substrates. The CVD system was assembled in a Lindberg-Blue horizontal furnace Model STF55433C with a 2-inch diameter quartz tube. All reactions were conducted at atmospheric pressure. A boat with 3.0 g of NH_4Cl was placed at the entrance of the furnace. The temperature at this position was maintained at 350°C, which is sufficient to sublime NH_4Cl and to promote the formation of a gaseous gallium chloride precursor [3, 4, 12, and 13]. A quartz crucible with 0.7 g of gallium was placed 10 cm downstream from the position of the ammonium salt and the sapphire substrate with the gold film, separated by 1.0 cm from the crucible with the gallium metal source. Unlike other reports, where the gold film has been annealed prior to the GaN synthesis to assure the presence of gold nanoparticles, in the experiment described herein , the substrate with the gold film was placed in the CVD reactor without annealing.

A schematic representation of this setup is shown elsewhere [4, 14]. The furnace was heated at a rate of 30°C·min^{-1} and maintained at 800°C. The tube was purged with ammonia when the temperature in the furnace was 300°C and the flow rate was constant at 180 sccm until the end of the reaction. Then, the system was cooled to room temperature by switching off the heater. The exhaust gases were collected in an aqueous HCl solution trap.

Characterization

Electron scanning microscopy images were collected with a JEOL JSM5300 system and a cathode luminescence (CL) system using an electron beam with an energy of 15 keV. The structures were analyzed with a Phillips X'pert-MDP diffractometer with Cu k radiation (0.15404 nm). Nanoparticle dimensions were estimated by analyzing SEM images with Image Tool v 3.0 [15]. TEM images were collected with a JEOL 2010 microscope and a FEI model TECNAI F-30 operated at 300 keV adapted with an EDS detector. Atomic Force Microscopy (AFM) images were acquired with a Nanoscope III system.

RESULTS AND DISCUSSION

SEM

The first step in the synthesis of GaN rods is the deposition of gold clusters onto the substrate surface. Instead of separately annealing the gold film prior to CVD synthesis [4], we prepared the gold clusters in situ by introducing the substrate with the gold film directly into the CVD reactor along with the reagents required for GaN deposition. This is the actual environment in which gold clusters promote the growth of GaN. This in situ annealing seemed feasible since the minimal temperature to form the clusters is 200°C, independent of time, which can range from 1 to 30 minutes [16]. Thus, by applying a heating rate of 30°C·min⁻¹, there is sufficient temperature and time (ca. 26 min) to form the clusters before reaching the operational temperature of 800°C for this reaction. A similar in situ annealing was successful when gold was deposited onto a silicon substrate [17]. Sapphire and quartz were used to verify influences of substrate crystallinity on either the formation of gold clusters or the growth of GaN.

GaN deposition only starts when the temperature in the furnace is 800°C, since this corresponds to 350°C at the entrance of the furnace where the NH_4Cl sublimes and transfer the metal gallium to the vapor phase [4, 18].

In two experiments, the substrates were removed at this step to analyze the clusters by XRD and SEM (Figures 1(a), (b)). The depositions carried out with quartz and sapphire substrates were identified by XRD as GaN with a hexagonal wurtzite-type structure (patterns not shown herein), matching with the JCPDS card 74-0243 [19]. The GaN deposited onto quartz exhibited relative intensities similar to that of the powder pattern of the card, whereas the sample grown onto sapphire showed greater intensity for the [002] reflection, as a consequence of the preferential orientation of the rods in relation to the substrate.

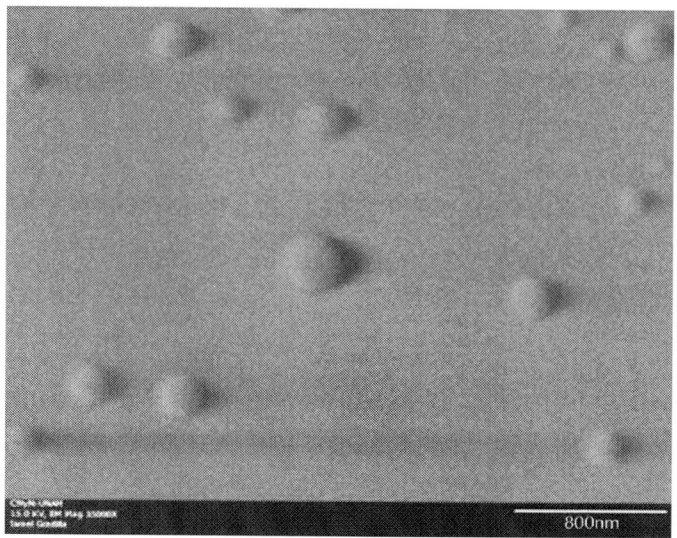

(a)

(b)

(c)

(d)

Figure 1: SEM images of gold clusters formed onto (a) sapphire and (b) quartz; and the rods grown with these clusters in (c) sapphire and (d) quartz.

Other substrates with gold films were treated with the same procedure and allowed to react 30 minutes at 800°C to obtain GaN rods (Figures 1(c) and (d)). The images were analyzed with an image program to determine particle dimensions [15].

It should be noted that due to the NH_3 atmosphere (applied when the furnace reached 300°C) and the reagents in the CVD reactor, traces of gallium, nitrogen or chloride could be present in the cluster. Henceforth, the clusters formed before the start of NH_4Cl sublimation will be called "clusters", while the particles which remained on the tips of GaN rods will be named "droplets" because they form liquid droplets during the growth of the GaN rods [6-8, 10].

Figure 2 shows the AFM image of gold clusters formed by annealing the gold film on sapphire. The sizes are ~250 nm, close to the results obtained from the analysis of SEM images.

Table 1 gives the mean dimensions of the clusters, rods, and droplets on the tips to indicate the relationship between these particles. The dimensions of initial gold clusters are essentially the same as the rods diameters on quartz (170 vs. 150 nm) and on sapphire (215 vs. 225 nm). These data suggest that the size of the pristine gold clusters (i.e., before GaN growth) may determine the diameter of the rods. Table 1 also shows that pristine clusters (before GaN growth) are slightly smaller than the droplets on the tip of the rods on quartz (170 vs. 210 nm) and sapphire (215 vs. 270 nm). This observation may be explained by swelling in the vapor-liquid (V-L) phase, where the gallium is dissolved by gold, thereby increasing the droplet size. Incorporation of nitrogen is discarded because of its negligible solubility in gold at the reaction temperature [8, 11].

10.00 nm

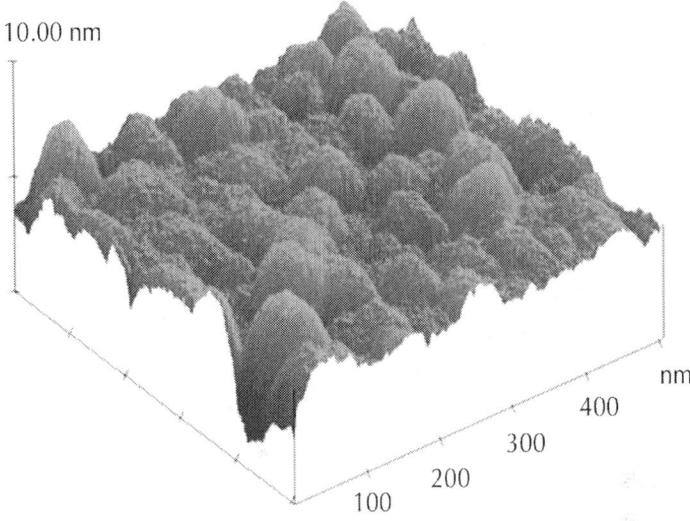

nm

400

300

200

100

Figure 2: AFM image of the gold particles formed on sapphire during the heating ramp of the CVD system.

Table 1: Relationship between sizes of gold clusters, droplets on the rod tips and rods

	Sapphire	Quartz
Initial clusters	215 nm (σ = 80)	170 nm (σ = 40)
Droplets on tips	270 nm (σ = 60)	210 nm (σ = 30)
Rod diameter	225 nm (σ = 40)	150 nm (σ = 50)

σ = Standard deviation calculated from 20 measurements.

To confirm the swelling of the cluster, commercial gold nanoparticles of ~50 nm in diameter were dropped onto sapphire and quartz substrates. These samples were then analyzed by SEM with back-scattered electrons to highlight the gold droplets (Figures 3(a) and (b)). The size of droplets without GaN formation is that of rod width (Figure 3(a)), suggesting that in the earlier stage of growth, the width depends on the cluster size. This confirms that the amount of gallium being dissolved by the droplet is transferred to the L-S interface. However, gallium diffusion through the droplet is slow and due to this, gallium concentrates itself, resulting in an increase in the size of the droplet. As a result, bigger droplet size is found particularly in large rods.

The presence of the droplet on the tip indicates that they may be responsible for the unidirectional growth of GaN. In principle, this observation could be associated with a vapor-liquid-solid (VLS) mechanism, as other authors have mentioned [8] and that the mechanism could be inferred from the morphology and size of the droplet as well as its relationship with the diameter of the rod [6]. This discussion is addressed in the next section.

TEM: Effects of Nitrogen Insolubility in the Droplet

TEM images of rods grown onto quartz and sapphire showed well-faceted droplets on the tips, which were also larger than the rod diameters. Figures 4(a) and 6(a) are examples of rods and droplets with these features which represent the majority of specimens observed. Nonetheless, isolated cases of quasi-spherical droplets were also found (Figures 4(b) and 6(b)).

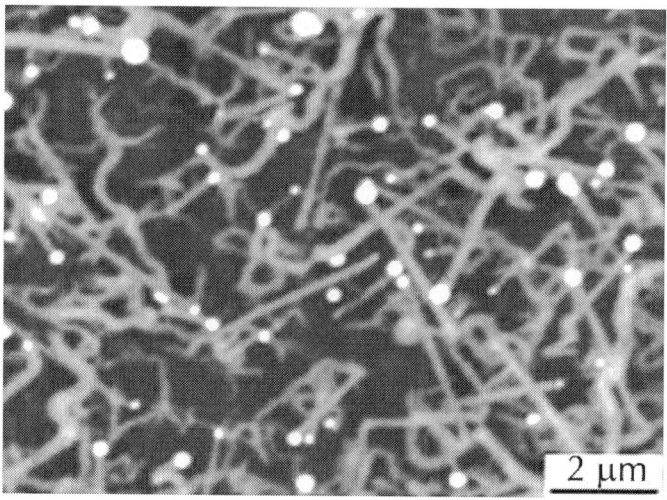

2 μm

(a)

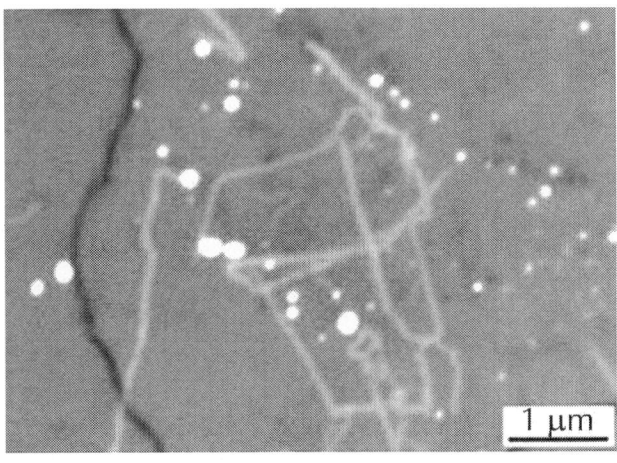

(b)

Figure 3: SEM images with back-scattered electrons of GaN rods and gold droplets onto (a) quartz and (b) sapphire.

A droplet in liquid state would be spherical, but when the temperature of the CVD system decreased the droplet crystallized with defined faces [8]. It is possible that the droplets were not only formed by gold because, as reaction centers for GaN growth, they should contain at least gallium or GaN. The spectra collected with the EDS detector of the TEM microscope show the profile of gallium and gold content along the length of a rod to the tip with the droplet (Figure 5). The intensity of the gallium spectrum decreased in the position from 0.5 to 0.15 µm, which could be a result of a gallium-deficient zone, as reflected in the cathode-luminescence spectrum that will be shown later. Then, when the electron beam reached the droplet (at 0.20 µm), the intensity for gold increased, whereas the signal for gallium decreased but did not disappear, confirming that the droplet is formed by a mixture of gold and gallium.

Although the presence of droplets on the rods is an indication that the growth followed the VLS mechanism [9], some variations could be involved. Since the elements forming the rods have to be dissolved by the metal catalyst [8], and nitrogen is not miscible in gold at 900°C [8, 11], the prediction indicates that GaN rods could not be grown by this method. However, the gold-catalyst did produce rods in the ammono-CVD system. In order to discount any possible unidirectional growth of GaN induced by the crystalline sapphire or the amorphous quartz, additional experiments performed in the absence of gold produced only continuous polycrystalline films [5], suggesting that the gold particles can act as a catalyst to form GaN rods under these experimental conditions.

The insolubility of nitrogen in the droplet seems to be unimportant for rod formation since the continuous flow of ammonia guarantees constant contact of nitrogen with the surface of the droplet containing gallium.

Thus, the contact of gaseous nitrogen with the liquid gold-gallium droplet seems to be sufficient to form GaN with a unidirectional structure as proposed by Gottschalch et al. [11]. Thus, the actual mechanism is probably related to a VLS mechanism with regards to gallium. Finally, the lack of nitrogen solubility in the droplet may account for the large number of crystalline faults along the rods, the irregular diameters, and the rough surfaces observed in these samples (Figures 4(a)-(d) and 6(a) - (c)).

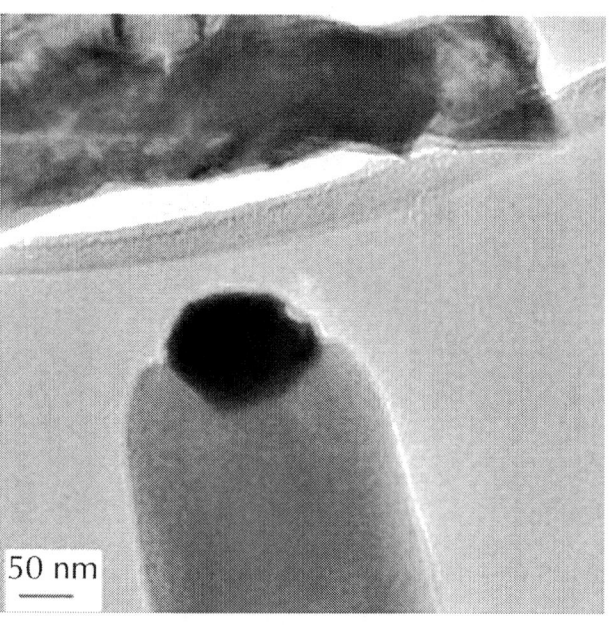

50 nm

(a)

50 nm

(b)

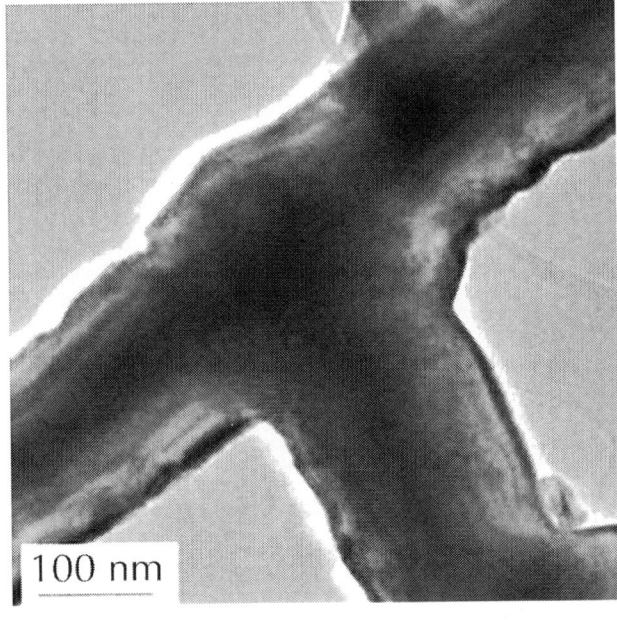

100 nm

(c)

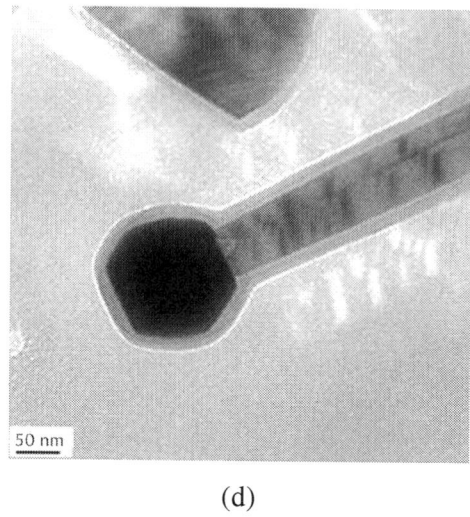

(d)

Figure 4: TEM images of GaN rods grown onto quartz.

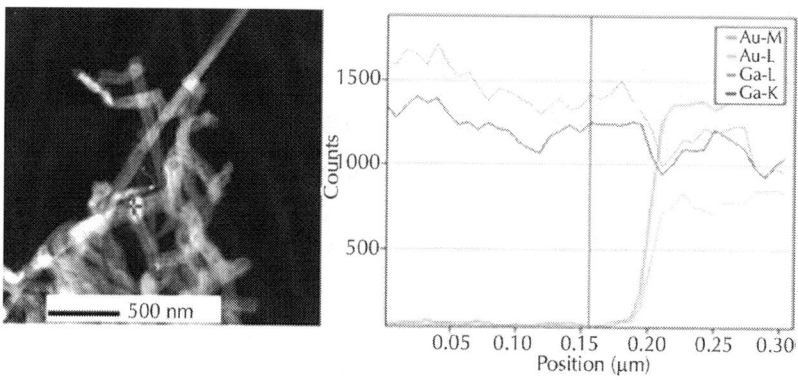

Figure 5: Gallium and gold content profile detected by EDS along the rod.

TEM: Effects of Gallium Diffusion through the Droplet and Gallium Concentration in the Gas Phase

The diffusion capacity of gallium through gold influences the morphology of the droplet (shape and size) and the rod [6].

The observations made by Mohammad [6] on rods with different compositions led to the conclusion that metal droplets are larger than the rod diameters because of low diffusion of the atoms that form the rod through the droplet in the liquid state. In our experiments, the droplets were clearly larger than the diameters of the rods, but in addition the droplets were well-faceted (Figures 4(a), (c) and 6(a)). If gallium diffuses at a slow rate through gold, it is expected that the droplet accumulates gallium and swells, and that larger droplets are then found at the end of the reaction in comparison to the size of the gold particles which are obtained by simple annealing, as shown in Figure 2.

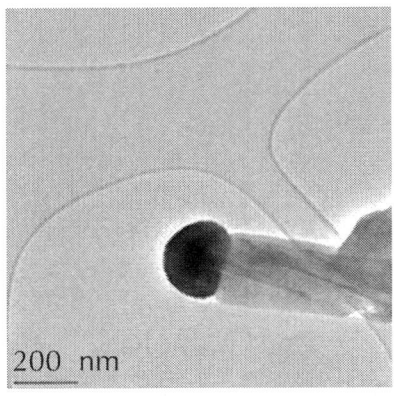

(a)

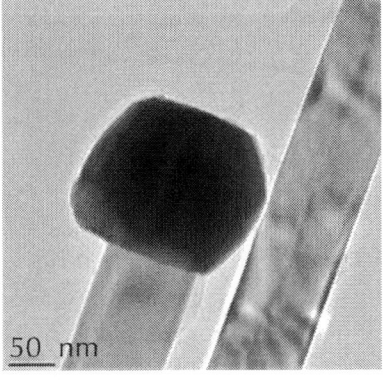

(b)

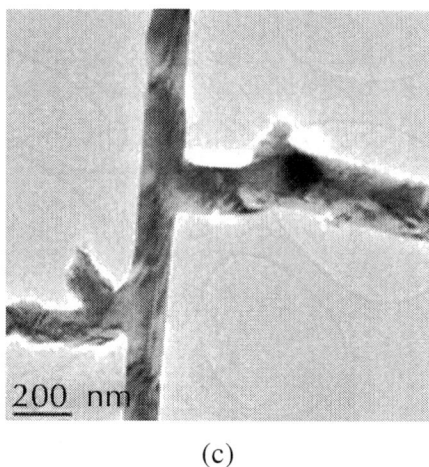

(c)

Figure 6: TEM images of GaN rods grown onto sapphire.

Another consequence of low diffusion of gallium could be poor transfer of gallium to the L-S interface, which could result in the formation of thin rods. On the contrary, the accumulation of gallium in the droplet could be responsible for the faceted morphology of the droplet, since gold nanoparticles of 200 nm tend to be spherical.

The widening of rods is inherent to this CVD system [5] and occurs in addition to the gold catalyzed growth. The surfaces of the GaN rods formed are continuously exposed to gallium and nitrogen reagents in the vapor phase in such a way that the vapor reagents could react on the road surface. One fact in accordance with this is found in the TEM images of rods onto quartz and sapphire. The size of droplets in Figures 4(a) and 6(a) are near to 100 nm, while the rods are thinner. However, even when the droplets in Figures 4(b) and 6(b) are still near to 100 nm, the width of the rods increased to 220 nm (Figure 4(b)) or to 100 nm (Figure 6(b)). This widening could not be related to the presence of gold but to the further reaction of the reagents in the vapor phase.

Another evidence of the extra widening was detected in the rods grown onto quartz. In this deposition there was a darker area visible at naked eye corresponding to the zone of the substrate exposed directly to the gallium boat. The SEM image of this region shows columnar structures (Figure 7) suggesting that the diameter increased by the side deposition of GaN, influenced by the exposition to a higher concentration of gallium reactive species.

TEM: Branches in GaN Rods

Another observation is that the rods formed branches with a diameter similar to or smaller than that of the main rods (Figures 4(d) and 6(c)). The orientations of branches did not follow a regular trend; they grew at angles ranging from 83 to 157 degrees, including T shaped branching.

The formation of these branches may be explained as secondary growth originating from the convergence of an L-S interface with the nitrogen present around the droplet in the V phase (Figure 8), which may react to form GaN by a V-S step.

As a result of the accumulation of gallium in the gold droplet (as discussed earlier), the possibilities to react with nitrogen increase, but solidification of GaN requires a L-S interface, which can be found close to the rod (dashed area in Figure 8). Evidences of incipient branching were found in samples of both substrates (Figures 9(a), (b)). The further growth of the branch cannot be associated with a gold catalyzed pathway because we did not find any droplet division. It was rather seen that the droplet either deviates and catalyzes the formation of a branch (Figure 6(c)), or follows the main rod to form a side branch (Figure 6(c)) according to the model depicted in Figure 8. Thus, the rod without the presence of a droplet should continue to grow by replicating the structure of crystalline GaN without the need for a catalyst.

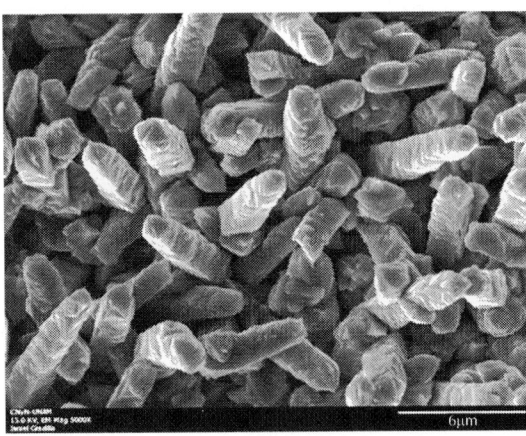

Figure 7: SEM image of the quartz zone exposed directly to the gallium stream.

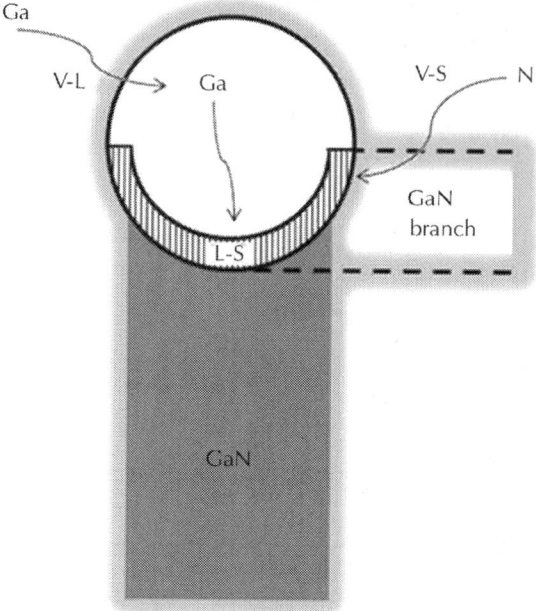

Figure 8: Scheme of GaN nucleation in the L-S interface.

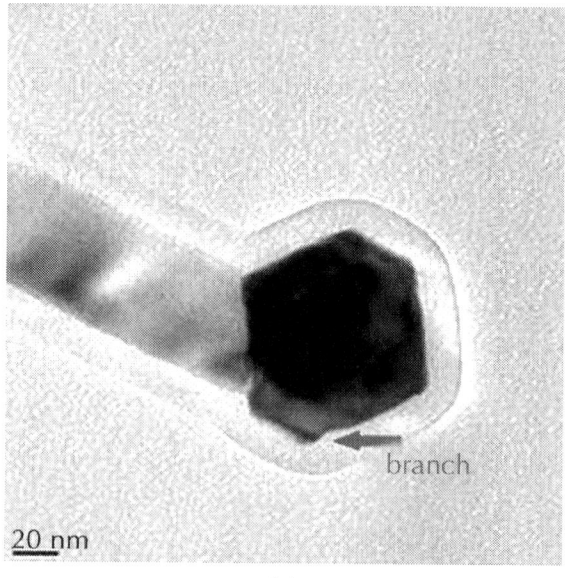

(a)

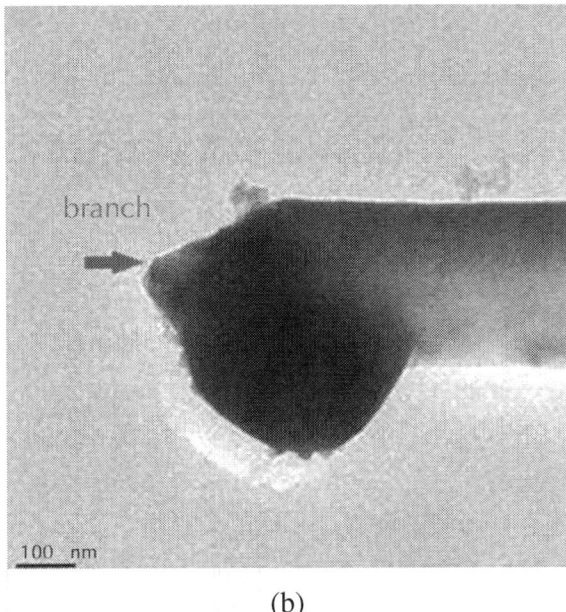

(b)

Figure 9: TEM images of GaN rods with incipient branching in samples grown onto (a) sapphire and (b) quartz.

HRTEM

The GaN rods contained a large amount of crystalline faults, as expected, since the gold droplet did not form a complete solution with nitrogen. Despite the faults and the large thickness of the rods, HRTEM images were collected which revealed that the rods grew along the c axis (Figure 10). Thus, the droplet is supported by the (0002) plane of GaN which has an interplanar spacing of 2.6 Å.

Cathodoluminescence

The CL spectra collected at room temperature (Figure 11) reveals a broad band-edge emission at 380 nm (3.26 eV) commonly associated with GaN with a wurtzite-like structure. Typical emission of the near band edge in GaN structures has been detected at 365 - 378 nm [20-22], which is the donor bound exciton emission [20]. This band can be broad [23] due to donor acceptor pair in GaN nanowires [20].

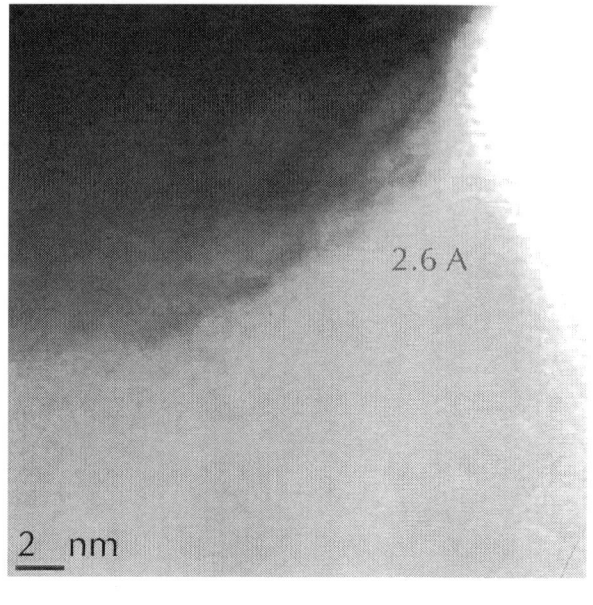

(a)

(b)

Figure 10: HRTEM images of GaN rod grown on (a) quartz and (b) sapphire.

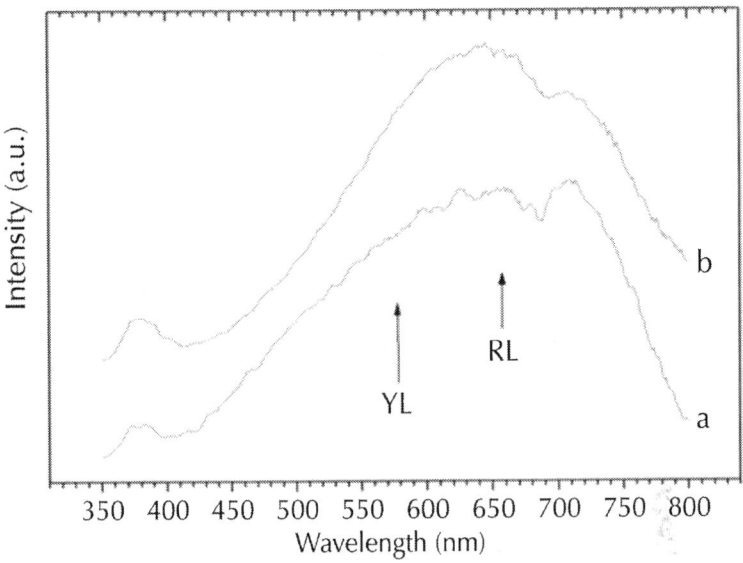

Figure 11: Room-temperature CL spectra of GaN deposited on (a) sapphire and (b) quartz.

In addition, a broad and high intensity band emission above 500 nm was observed and assigned to defect-rich crystallites [24, 25]. The broad band is composed of yellow luminescence (YL) and red luminescence (RL). The former is caused by gallium vacancies while the red luminescence is produced in regions with nitrogen deficiency [22]. Both of these regions are exhibited by the GaN rods obtained on sapphire and quartz, indicating that the formation of defects depends on the type of reagents used in the ammono-CVD system and not on the substrate.

The RL intensity was even higher than the YL, thus indicating several sites in the GaN structure with nitrogen deficiency. This fact has coherency with the hypothesis stating that nitrogen is poorly soluble in the gold droplet, i.e. the lower possibility of nitrogen to reach the L-S interface, in comparison with gallium, produces nitrogen deficient sites. The presence of crystalline defects makes GaN prone to react with some chemical species like gases or even with more complex molecules like amino acids as some preliminary experiments proved, and such surface reactions induce changes in electrical conductivity. This behavior can be useful to design chemical sensors. Additionally, the 1D structures obtained in this work provide a larger contact area in comparison with films.

CONCLUSIONS

Formation of gold clusters in situ is feasible during the heating ramp of the CVD system with a heating rate of $30°C·min^{-1}$ up to $800°C$. Gold particles formed onto amorphous or crystalline surfaces like quartz and sapphire substrates are effective in catalyzing the growth of GaN rods. The diameter of the rods depends on the size of the gold particles, without the influence of the sapphire or quartz substrate surface once the growth of crystalline GaN had been initiated.

GaN rods grow by a VLS mechanism even when only gallium is dissolved by the droplet, by reaction in the V-S region with nitrogen. Since nitrogen surrounds the gallium/gold droplet surface, the region near the L-S interface could form branches; when this occurred, the droplet is not divided and guides either the main rod or a branch. The rods undergo lateral growth, influenced by the gallium complex in the vapor phase. Gallium has a low rate of diffusion through the gold droplet, resulting in the accumulation of gallium, thereby increasing the volume of the droplet. After cooling, well-faceted droplets formed on the tips. The GaN on both substrates possessed a wurtzite-like structure according to the X-ray analysis and CL spectra.

Owing to the insolubility of nitrogen in the droplet, rods exhibit crystalline defects, reflected in the wide red band of the CL spectra, despite this, the crystalline defects represent an opportunity to develop devices for gas sensing since this sites are prone to react with foreign chemical compounds changing the electrical conductivity. The use of NH_4Cl to transfer gallium to the vapor phase is a good alternative to prepare GaN in a CVD system, being possible to grow rods with gold as catalyst.

ACKNOWLEDGEMENTS

We are grateful to the technical assistance of I. Gradilla, E. Aparicio, F. Ruiz, E. Flores, J.A. Diaz Hernandez, D. Dominguez and M.I. Perez Montfort. Financial support was provided by DGAPA-UNAM (project

IN109612) and CONACYT (project 82984). We also thank M. Avalos Borja for helpful suggestions and for providing access to LINAN's electron microscopy facilities.

REFERENCES

1. E. Estephan, et al., "Tailoring GaN Semiconductor Surfaces with Biomolecules," Journal of Physical Chemistry B, Vol. 112, No. 29, 2008, pp. 8799-8805. doi:10.1021/jp804112y

2. M. He, et al., "Growth of GaN Nanowires by Direct Reaction of Ga with NH_3," Journal of Crystal Growth, Vol. 231, No. 3, 2001, pp. 357-365. doi:10.1016/S0022-0248(01)01466-X

3. D. Ehrentraut, et al., "Physico-Chemical Features of the Acid Ammonothermal Growth of GaN," Journal of Crystal Growth, Vol. 310, No. 5, 2008, pp. 891-895.doi:10.1016/j.jcrysgro.2007.11.090

4. R. Garcia, A. C. Thomas and F. Ponce, "Measurement of the Solubility of Ammonia and Nitrogen in Gallium at Atmospheric Pressure," Journal of Crystal Growth, Vol. 467, No. 1-2, 2008, pp. 3131-3134. doi:10.1016/j.jcrysgro.2008.03.030

5. G. G. C. Arízaga, et al., "Influence of Reaction Conditions on the Growth of GaN Rods in an Ammono-CVD Reactor," Journal of Crystal Growth, Vol. 319, No. 1, 2011, pp. 19-24.doi:10.1016/j.jcrysgro.2011.01.103

6. S. N. Mohammad, "Why Droplet Dimension Can Be Larger than, Equal to, or Smaller than the Nanowire Dimension," Journal of Applied Physics, Vol. 106, 2009, Article ID: 104311, pp. 1-11.

7. A. Morales and C. M. Lieber, "A Laser Ablation Method for the Synthesis of Crystalline Semiconductor Nanowires," Science, Vol. 279, No. 5348, 1998, pp. 208-211.doi:10.1126/science.279.5348.208

8. X. Duan and C. M. Lieber, "Laser-Assisted Catalytic Growth of Single Crystal GaN Nanowires," Journal of the American Chemical Society, Vol. 122, No. 1, 2000, pp. 188-189. doi:10.1021/ja993713u

9. L. Yu, Y. Ma and Z. Hu, "Low-Temperature CVD Synthesis Route to GaN Nanowires on Silicon Substrate," Journal of Crystal

Growth, Vol. 310, No. 24, 2008, pp. 5237-5240.doi:10.1016/j. jcrysgro.2008.09.191

10. R. S. Wagner and W. C. Ellis, "Vapor-Liquid-Solid Mechanism of Single Crystal Growth," Applied Physics Letters, Vol. 4, No. 89, 1964, pp. 89-90. doi:10.1063/1.1753975

11. V. Gottschalch, et al., "VLS Growth of GaN Nanowires on Various Substrates," Journal of Crystal Growth, Vol. 310, No. 23, 2008, pp. 5123-5128.doi:10.1016/j.jcrysgro.2008.08.013

12. P. Purdy, "Ammonothermal Synthesis of Cubic Gallium Nitride," Chemistry of Materials, Vol. 11, No. 7, 1999, pp. 1648-1651. doi:10.1021/cm9901111

13. T. Hashimoto, et al., "Growth of Gallium Nitride via Fluid Transport in Supercritical Ammonia," Journal of Crystal Growth, Vol. 275, No. 1-2, 2005, pp. e525-e530.doi:10.1016/j. jcrysgro.2004.11.024

14. G. G. C. Arizaga, et al., "Reversible Intercalation of Ammonia Molecules into a Layered Double Hydroxide Structure without Exchanging Nitrate Counter-Ions," Journal of Solid State Chemistry, Vol. 183, No. 10, 2010, pp. 2324-2328. doi:10.1016/j. jssc.2010.07.050

15. Image Tool Version 3.0. http://ddsdx.uthscsa.edu/dig/itdesc.html

16. Ch. Y. Chang, et al., "Control of Nucleation Site Density of GaN Nanowires," Applied Surface Science, Vol. 253, No. 6, 2007, pp. 3196-3200.doi:10.1016/j.apsusc.2006.07.007

17. Ch. Cao, X. Xiang and H. Zhu, "High-Density, Uniform Gallium Nitride Nanorods Grown on Au-Coated Silicon Substrate," Journal of Crystal Growth, Vol. 273, No. 3-4, 2005, pp. 375-380. doi:10.1016/j.jcrysgro.2004.09.050

18. S. E. Alexandrov, A. Y. Kovalginy and D. M. Krasovitskiy, "A Study of CVD of Gallium Nitride Films by In-Situ Gas-Phase UV Spectroscopy," Journal de Physique IV, Vol. 5, No. C5, 1995, pp. 183-190. doi:10.1051/jphyscol:1995520

19. Data Collection of the Joint Committee on Powder Diffraction Standard, PCPDFWIN Version 2.2, June 2001.

20. Y. H. Ra, et al., "The Influence of the Working Pressure on the Synthesis of GaN Nanowires by Using MOCVD," Journal of Crystal

Growth, Vol. 312, No. 6, 2010, pp. 770-774. doi:10.1016/j.jcrysgro.2009.12.056

21.   R. Navamathavan, et al., "Different Growth Behaviors of GaN Nanowires Grown with Au Catalyst and Au + Ga Solid Solution Nano-Droplets on Si(111) Substrates by Using MOCVD," Current Applied Physics, Vol. 11, No. 1, 2011, pp. 77-81.doi:10.1016/j.cap.2010.06.022

22.   D. S. Chander, J. Ramkumar and S. Dhamodaran, "Controlled 1-D to 3-D Growth Mode Transition of GaN Nanostructures and Their Optical Properties," Physica E, Vol. 4. No. 9, 2011, pp. 1683-1687. doi:10.1016/j.physe.2011.05.022

23.   Y. H. Cho, et al., "Optical Properties of Laterally Overgrown GaN Pyramids Grown on (111) Silicon Substrate," Current Applied Physics, Vol. 2, No. 6, 2002, pp. 515- 519.doi:10.1016/S1567-1739(02)00168-2

24.   S. Q. Zhou, et al., "Comparison of the Properties of GaN Grown on Complex Si-Based Structures," Applied Physics Letters, Vol. 86, No. 8, 2005, pp. 1-3.doi:10.1063/1.1868870

25.   A. N. Red'kin, et al., "Chemical Vapor Deposition of GaN from Gallium and Ammonium Chloride," Inorganic Materials, Vol. 40, No. 10, 2004, pp. 1049-1053.doi:10.1023/B:INMA.0000046466.62619.e9

10

Selection Method of Surfactants for Chemical Enhanced Oil Recovery

Roland Nagy[1], Rubina Sallai[1], László Bartha[1], and
Árpád Vágó[2]

[1]Department of MOL-Hydrocarbon and Coal Processing, University of
Pannonia, Veszprém, Hungary
[2]MOL Plc. Research and Business Development, Budapest, Hungary

ABSTRACT

Alternative energy sources have not yet presented suitable to meet the energy demand of the world; therefore crude oil will play furthermore an essential role in the energy consumption in the future. EOR is a challenging field for several scientific disciplines. The number of patents highlights the importance of this area. Most of the publications label that the target of the chemical processes is the reduction of IFT between the displacing liquid and oil phase. Based on the results in the last two decades the surfactants and flow modifier type polymers have shown more potential for a higher efficiency of the EOR than in any other methods. The aim of this work was to develop different surfactant testing methods that are capable to characterize the most important surfactant properties separately and to evaluate their combined or complementary effects. There was no correlation between the specific characteristics. But the weak correlation was detected by using the CE complex evaluation method. Further improvements could contribute significantly to improve the selectivity of the new experimental non-ionic surfactants for EOR applications.

INTRODUCTION

Alternative energy sources have not yet presented suitable to meet the energy demand of the world; therefore crude oil will play furthermore an essential role in the energy consumption in the future. Considering the fact that the easily recoverable oil is running out and much oil remains in the reservoir after conventional methods have been exhausted, the implementation of Enhanced Oil Recovery has become crucial to guarantee a continuing crude oil supply [1].

The Enhanced Oil Recovery (EOR) technology covers the injection of specific type of a fluid or fluids into the reservoir by several methods (e.g.: chemical, thermal and microbial). The injected fluid promotes to dislocate of crude oil toward the producing well. Besides, the injected fluids interact with the reservoir rock/oil system and generate advantageous conditions for oil recovery. These interactions incorporate lowering the interfacial tension (IFT), improving the flow properties, modify wettability and help developing preferential phase behaviour. As a consequence of the interactions, physical and chemical mechanisms can occur, as well as the formation of thermal energy [2] - [5].

EOR is a challenging field for several scientific disciplines. The number of patents highlights the importance of this area. Most of the publications label that the target of the chemical processes is the reduction of IFT between the displacing liquid and oil phase. Based on the results in the last two decades the surfactants and flow modifier type polymers have shown more potential for a higher efficiency of the EOR than in any other methods [6] [7] .

Chemical flooding of oil reservoirs could be the one of the most successful method to increase oil recovery rate of the depleted reservoirs. The research and developing projects of the EOR surfactants are very costly due to the expensive field tests. Numerous screening test methods have been elaborated to reduce the costs and to estimate the potential efficiency of the tested surfactant compositions [3] [7] [8].

However, the previous experiments showed that the surfactant composition has to be effective under different and often extreme conditions, which are complicated to be modelled in the laboratories. There is no generally accepted test method for the selection of surfactants. Therefore, the further development of screening methods remains important.

The aim of this work was to develop different surfactant testing methods that are capable to characterize the most important surfactant properties separately and to evaluate their combined or complementary effects.

MATERIALS

Preparation of the Surfactant and Surfactant-Polymer Solutions

For the preparation of the experimental colloid solutions one anionic (MOLANIONIC) and two nonionic (PENONIONIC I and II) surfactants were used, that were developed and produced by cooperative research groups of University of Pannonia and Hungarian Oil and Gas Company (MOL Plc). Furthermore partially hydrolysed synthetic polyacrylamide was applied as a flow modifier type polymer.

The polymer-surfactant solutions were prepared by using brine, derived from the petroleum reservoir of Algy (Hungary).

The surfactant packages were prepared of the following components (detailed in Table 1).

The flow modifier type polymer (PAM type sulfonated co-polymer) was used in a concentration of 1000 ppm. The total concentration of the surfactant packages in the mixtures was 15,000 ppm.

METHODS

A complex laboratory screening method was developed to definite the properties of EOR type surfactants by the application of the following methods to determine the surface-active properties and also to study the possibility of the complex evaluation of the measurement data.

The Investigation of the Emulsifying Efficiency (EE)

In this method first the 50% of brine containing 1.5% of the emulsifier packages and 0.1% polymer and 50% of model oil (crude oil) should be shaken together at room temperature over 7 complete translation cycles. After 30 minutes long storage the volume of the formed phases should be measured. Thereafter 4 hours of storage at 80°C the volumes of the various phases should be determined again. The amount of aqueous phase and the emulsion- and oil phases are given in volume% related to the total liquid volume.

The maximum difference between the parallel results was in the interval of ±2%.

Table 1: Composition of the surfactants and surfactant-polymer mixtures

Symbols of package compositions	ST-marked mixtures	SM-marked mixtures
Components	Ratio of components, m/m%	
MOLANIONIC	55	60
PENONIONIC I	25	40
PENONIONIC II	20	-

Determination of the Oil Displacement Effect (ODE) By Thin Layer Chromatography (TLC)

The complex surface activity method has not been reported in the specific publications, which could be the ex- tent to porous surfaces characterized by washing efficiency of the oil. It was developed a method for oil flushing effect, that based on the foundations of thin layer chromatography of organic chemical analysis has been used.

The first step of the test procedure is the preparation of the thin layer. First the glass slab should be washed with water and n-heptane and let dry. Thereafter the slab should be plunged into chloroform-real core sand dis- persion and let dry for 20 minutes. Then 4 µl of crude oil should be spotted on the thin layer at 2 cm from the base of the glass slab. Then the sample should be run by pH adjusted (between 8.0 - 8.5) brine containing 1% of the emulsifier mixture at 80°C for 3 hours. The drying of the thin layer should be carried out at room tempera- ture after 24-hour storage. The oil displacement efficiency (ODE) of the surfactant composition can be evaluated by the distance from the centre of the oil spot to the upper edge of the spot measured in mm. The layers after the 3-hours run-time should be photographed with a digital camera, and the data should be recorded and stored in a database.

The maximum difference between the parallel results was in the interval of ±10%.

Determination of Interfacial Tension (IFT)

The spinning drop method (rotating drop method) is one of the methods used to measure the oil-water interfacial tension. The measurements should be carried out in a rotating horizontal tube, which contains a dense fluid, and drop of a less dense liquid should be placed inside the fluid. Till the rotation of the horizontal tube creates a centrifugal force towards the tube walls, the liquid drop will start to deform into an elongated shape; this elongation stops when the interfacial tension and centrifugal forces are balanced. The surface tension between the two liquids (for bubbles: between the fluid and the gas) can then be derived from the shape of the drop at this equilibrium point. A device used for such measurements is called a "spinning drop tensiometer".

The spinning drop method is usually preferred for the accurate measurements of surface tensions below 10^{-2} mN/m. It refers to either using the fluids with low interfacial tension or working at very high angular velocities.

The maximum difference between the parallel results was in the interval of ±5%.

Oil Displacement Test (ODT)

The special core flooding tests were performed by a test method simulated the processes in the high-temperature oil reservoirs. The crude oil was originated from Algy , Hungary . Testing temperature was 80°C. The tests were performed in a core with inner diameter of 2.5 cm and length of 30 cm. The heterogeneousness (homogeneity) of the formation was simulated by consolidated core sample.

Surfactant-polymer flooding has been successfully used to recover waterflood residual oil with acidic crude oils in laboratory-scale experiments. The additional oil recovery was determined, that was related to residual oil saturation after the water flooding [8].

The maximum difference between the parallel results was in the interval of ±5%.

RESULTS AND DISCUSSION

The first aim in this work was to investigate how and by which selectivity can be used the laboratory test methods for characterization and evaluation of the experimental surfactant mixtures for EOR purposes.

In our study, as a model two different packages of experimental surfactants were examined (of two-and three-components). The suitability of these surfactant packages for EOR purposes has been studied by the presented test methods.

The three screening methods and the oil displacement test were used also for the complete classification of surfactant packages. The second aim of this work was to use the screening methods for the optimization the compositions of two surfactant packages consisting non-ionic type compounds with different molecular structure by variation of the ratio of the polar and nonpolar groups in the mixture of molecules obtained by reaction of fatty acids and a polyalkanol-amine.

Investigation of SM-Surfactants Mixtures (Sequence of Two-Component)

The polymer-surfactants mixtures should be used in a reservoir, where the temperature is between 80°C - 100°C. Therefore it was important to study the relationship between the high temperature surface-active properties and the composition of the non-ionic components used in the surfactant mixtures. The data of the SM-surfactants series are given in Table 2 and demonstrated in Figures 1-3.

Generally the additional oil yield in the Enhanced Oil Recovery is related to initial stock. In these experiments and after the water flooding the residual oil contents were between 40% - 60%.

The IFT region was published by the literature as an essential condition for EOR application, but the figure shows that there is no correlation between the displacement results and IFT values. However, previous res- earches proved, the IFT values below 0.075 mN/m were advantageous for Enhanced Oil Recovery application, what criteria related to each sample had satisfied [8].

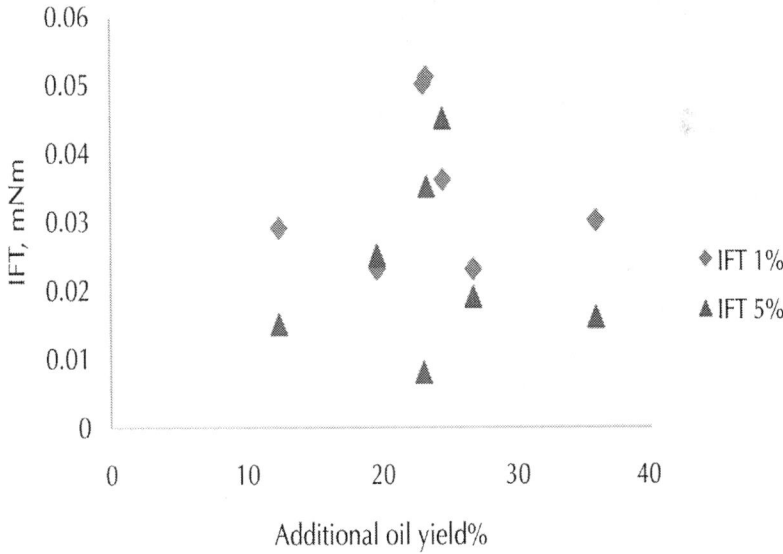

Figure 1: Relationship between the additional oil yield and the IFT.

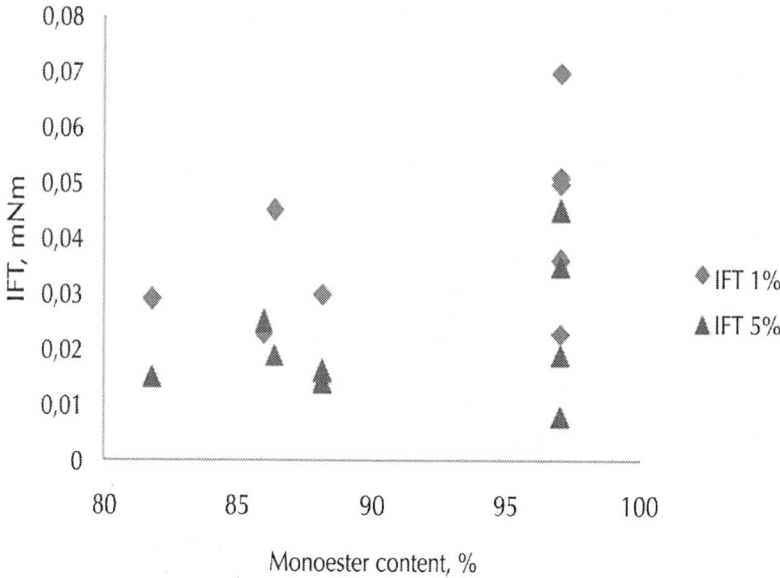

Figure 2: Relationship between the monoester content and IFT.

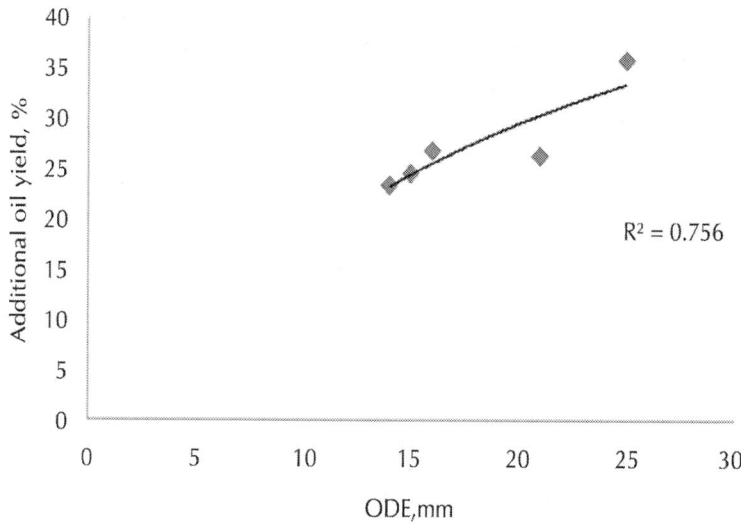

Figure 3: Relationship between the ODE and additional oil yield.

Table 2: Data of the surfactants and surfactant-polymer mixtures

Composition of the nonionic surfactant (PENONIONIC-1)			IFT, mNm		EE	ODE	Additional oil yield
	Monoester, %	Diester, %	1%	5%	1h, 80°C, %	mm	%
SM-1	88.17	4.1	0.03	0.016	100	25	35.94
SM-2	97.00	2.46	0.05	0.008	97.5	20	23.20
SM-3	81.85	16.9	0.029	0.015	10	15	12.45
SM-4	86.00	13.37	0.023	0.025	10	21	19.72
SM-5	78.93	15.12	0.039	0.028	50	16	-
SM-6	86.39	12.21	0.045	0.019	12.5	19	-
SM-7	97.00	2.46	0.051	0.035	95	14	23.42
SM-8	97.00	2.46	0.036	0.045	85	15	24.64
SM-9	97.00	2.46	0.023	0.019	40	16	26.89

Between the monoester content and interfacial tension there was not clear tendency as well. The other presented methods (EE, ODE) are mainly effect on the mobility properties of the oil/water emulsion, which have been found appropriate in the measured range. Experimental results of each test methods were not correlated with the replacement characteristics.

It was found, if the additional oil yield values are over 20%, than it can be evaluated as a successful recovery composition. Between the ODE and additional oil field a saturation curve type tendency was observed.

The additional oil yield method is considered to be the most important method of classification of surfactants for EOR. Therefore based on these results the ODE method could be also suitable and significantly cheap and fast method in the pre-selection process of the potential EOR surfactants

The higher diester content seemed advantageous to achieve an extreme low IFT value. Therefore the correlation could be used also for optimization of the composition of the non-ionic surfactant packages.

Investigation of ST-Surfactants Mixtures (Sequence of Three-Component)

The data of the ST-surfactants are given in Table 3. And demonstrated in Figure 4 and Figure 5. According to the preliminary tests the ST surfactant mixtures seemed also appropriate for EOR purposes that were supported by numerous of measurement data.

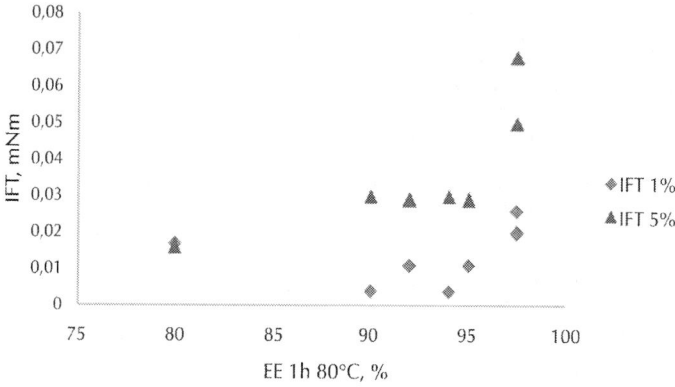

Figure 4: Relationship between the emulsifying effect and IFT.

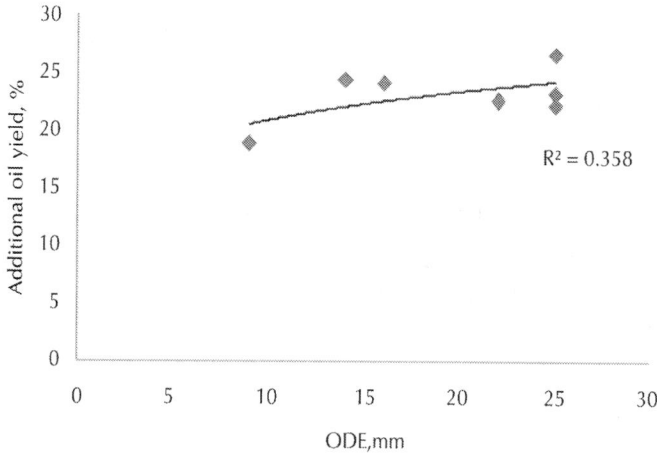

Figure 5: Relationship between the ODE and additional oil yield.

Table 3: Data of the surfactants and surfactant-polymer mixtures

Composition of the nonionic surfactant (PENONIONIC-1)		IFT, mNm		EE	ODE	Additional oil yield
Monoester, %	Diester, %	IFT 1%	IFT 5%	1h 80°C, %	mm	%
ST-1	97.00	2.46	0.004	0.03	90	16	24.03
ST-2	97.00	2.46	0.026	0.068	97.5	9	18.08
ST-3	86.39	12.21	0.011	0.029	95	25	22.13
ST-4	86.39	12.21	0.02	0.1	97.5	26	23.09
ST-5	81.85	16.9	0.011	0.029	92	22	22.46
ST-6	86.00	13.57	0.017	0.016	80	14	24.35
ST-7	78.93	15.12	0.004	0.013	94	25	26.50

The IFT values were found below 0.1 m · Nm, therefore this series of the surfactant mixtures can be suitable as well for EOR application. Between the emulsifying effect and the interfacial tension there was not clear tendency too. It was also found, that the IFT method in itself is not enough for classification of surfactants for EOR.

On the other hand it was found, that the additional oil yield values were also over 20% that can be considered a successful oil recovery ratio. In addition between the ODE and additional oil yield values an increasing tendency was observed with a low level of significance.

CE Calculation

Based on the results of the single laboratory test methods it was supposed that the high replacement efficiency of the surfactant packages only by a well balanced combination of the surfactant properties can be achieved. That is why a complex evaluation system was introduced and studied.

For the combined evaluation (CE) of the three properties an evaluation-equation was created and used to characterize the potential efficiency of the experimental surfactant for EOR. According to the mathematical calculations by using the measurement the best approximated equation was defined by the next formula:

$$CE = \frac{ODE \cdot 0.25 \cdot EE}{IEF^2 \cdot 100}$$

The calculated CE value of the series of the SM and ST surfactants in a function of additional oil yield related to residual oil shows Figure 6.

It was concluded, that by increasing the CE value a better additional oil yield could be estimated. In spite of the weak correlation it is supposed that by using also other complementary characteristics of the surfactant packages the estimation of oil recovery may be further improved. Figure 6 showed that in accordance with the general practice by using these experimental surfactant mixtures only a limited value (80%) from the residual crude oil can be recovered by this SP method. Based on these results next enhancements are planned, which could further contribute to reduce the selection of new surfactants. By the use of the developed methods the number of performed measurement can be approximately one-tenth reduced.

CONCLUSIONS

The experimental results were summarized as follows:

- A new complex laboratory method was developed to estimate the efficiency of the experimental surfactant packages for EOR.
- Different levels of correlations have been obtained between the surface-active properties and the additional oil yield.
- Slightly correlation was found between the CE factor and the additional oil yield that could open new opportunities for cost reduction of the development.
- By these measurements the number of experimental surfactants and their mixtures can be reduced.

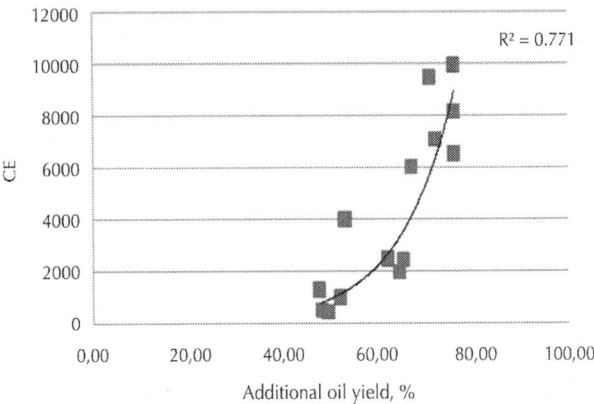

Figure 6: Relationship between the CE and additional oil yield (related to residual oil) in case of SM and ST-surfactants.

There was no correlation between the specific characteristics. But the weak correlation was detected by using the CE complex evaluation method. Further improvements could contribute significantly to improve the selectivity of the new experimental non-ionic surfactants for EOR applications.

REFERENCES

1.   Thomas, S. (2008) Enhanced Oil Recovery—An Overview. Oil & Gas Science and Technology, 63, 9-19. http://dx.doi.org/10.2516/ogst:2007060.

2.   Gaillard, N., Giovannetti, B. and Favero, C. (2010) Improved Oil Recovery Using Thermally and Chemically Protected Compositions Based on Co- and Terpolymers Containing Acrylamide. SPE Improved Oil Recovery Symposium, Tulsa, 24-28 April 2010, Document ID: SPE-129756-MS. http://dx.doi.org/10.2118/129756-MS.

3.   Levitt, D.B. and Pope, G.A. (2008) Selection and Screening of Polymers for Enhanced-Oil Recovery. SPE Symposium on Improved Oil Recovery, Tulsa, 20-23 April 2008, Document ID: SPE-113845-MS.

4.   Zhong, C, Luo, P., Ye, Z. and Chen, H. (2009) Characterization and Solution Properties of a Novel Water-Soluble Terpolymer for Enhanced Oil Recovery. Polymer Bulletin, 62, 79-89.http://dx.doi.org/10.1007/s00289-008-1007-6.

5.   Schramm, L.L. (2000) Surfactants: Fundamentals and Applications in the Petroleum Industry. Cambridge University Press, Cambridge.

6.   Myers, D. (2006) Surfactant Science and Technology. Wiley-Interscience, Hoboken.

7.   Jeirani, Z., Mohamed Jan, B., Si Ali, B., Noor, I.M., See, C.H. and Saphanuchart, W. (2013) Formulation, Optimization and Application of Triglyceride Microemulsion in Enhanced Oil Recovery. Industrial Crops and Products, 43, 6-14.http://dx.doi.org/10.1016/j.indcrop.2012.07.002.

8.   Nasr-El-Din, H.A., Hawkins, B.F. and Green, K.A. (1992) Recovery of Residual Oil Using the Alkali/Surfactant/ Polymer Process: Effect of Alkali Concentration. Journal of Petroleum Science and Engineering, 6, 381-401. http://dx.doi.org/10.1016/0920-4105(92)90064-8.

Citations

CHAPTER 1

M. Opgenorth, W. McDermott, P. Laz and C. Lengsfeld, "A Combined Probabilistic and Optimization Approach for Improved Chemical Mixing Systems Design," Engineering, Vol. 3 No. 6, 2011, pp. 643-652. doi:10.4236/eng.2011.36077.

CHAPTER 2

Kris Lawry and Dirk John Pons, "Integrative Approach to the Plant Commissioning Process," Journal of Industrial Engineering, vol. 2013, Article ID 572072, 12 pages, 2013, doi:10.1155/2013/572072.

CHAPTER 3

Bat-Erdene, E., Byambagar, B., Enkhtsetseg, E. and Avid, B. (2014) Chemical Analysis on Mongolia's Natural Bitumen. Advances in Chemical Engineering and Science, 4, 184-188. doi: 10.4236/aces.2014.42021.

CHAPTER 4

Shirish H. Sonawane, Sarang P. Gumfekar, Kunal H. Kate, et al., "Hydrodynamic Cavitation-Assisted Synthesis of Nanocalcite," International Journal of Chemical Engineering, vol. 2010, Article ID 242963, 8 pages, 2010. doi:10.1155/2010/242963.

CHAPTER 5

A. Cunha, M. Pacheco and J. Bergmann, "Influence of the Chemical Composition of Completion Fluids on the Propagation of Electromagnetic Waves within Oil Wells," Engineering, Vol. 4 No. 12A, 2012, pp. 966-971. doi:10.4236/eng.2012.412A122.

CHAPTER 6

F. Mohammed Nasser El-Fayez, "Effects of Chemical Reaction on the Unsteady Free Convection Flow past an Infinite Vertical Permeable Moving Plate with Variable Temperature," Journal of Surface Engineered Materials and Advanced Technology, Vol. 2 No. 2, 2012, pp. 100-109. doi: 10.4236/jsemat.2012.22016.

CHAPTER 7

Hinkov, I., Farhat, S., Lungu, C., Gicquel, A., Silva, F., Mesbahi, A., Brinza, O., Porosnicu, C. and Anghel, A. (2014) Microwave Plasma Enhanced Chemical Vapor Deposition of Carbon Nanotubes. Journal of Surface Engineered Materials and Advanced Technology, 4, 196-209. doi: 10.4236/jsemat.2014.44023.

CHAPTER 8

A. Benosman, M. Mouli, H. Taibi, M. Belbachir, Y. Senhadji, I. Bahlouli and D. Houivet, "Studies on Chemical Resistance of PET-Mortar Composites: Microstructure and Phase Composition Changes," Engineering, Vol. 5 No. 4, 2013, pp. 359-378. doi: 10.4236/eng.2013.54049.

CHAPTER 9

G. Guadalupe Carbajal Arízaga, K. Viridiana Chávez Hernández, N. Cayetano Castro, M. Herrera Zaldivar, R. García Gutiérrez and O. Edel Contreras López, "Synthesis and Characterization of GaN Rods Prepared by Ammono-Chemical Vapor Deposition," Advances in Chemical Engineering and Science, Vol. 2 No. 2, 2012, pp. 292-299. doi:10.4236/aces.2012.22034.

CHAPTER 10

Nagy, R., Sallai, R. , Bartha, L. and Vágó, Á. (2015) Selection Method of Surfactants for Chemical Enhanced Oil Recovery. Advances in Chemical Engineering and Science, 5, 121-128. doi: 10.4236/aces.2015.52013.

Index

A

Advanced Mean Value (AMV) 6
Atomic force microscopy (AFM)
    129

C

Cathode luminescence (CL) 194
Chemical reaction 95, 96, 97,
    98, 99, 104
Chemical vapor deposition
    (CVD) 116, 192
Combined evaluation (CE) 227
Computational fluid dynamics
    (CFD) 2, 116, 120

D

Differential scanning calorimetry
    (DSC) 144
Differential thermal analysis
    (DTA) 144

E

Electrical discharge machine
    (EDM) 18
Emulsifying Efficiency (EE) 220
Enhanced Oil Recovery (EOR)
    218

F

Foreign element catalytic agent
    (FECA) 193

G

Genetic Algorithms (GA) 2

H

Hydrogen-Iodide Chlorine (HICl)
    8

I

Incorporate lowering the interfa-
    cial tension (IFT) 218
Integration definition zero
    (IDEF0) 32
Interfacial Tension (IFT) 221

L

Local Thermodynamic Equilib-
    rium (LTE) 121

M

Magnetohydrodynamic (MHD)
    97
Mean-value (MV) 6
Multi-walled carbon nanotubes
    (MWCNTs) 115

N

New Zealand Aluminium Smelt-
    ers (NZAS) 40

O

Oil Displacement Effect (ODE)
    221
Oil displacement efficiency
    (ODE) 221
Oil Displacement Test (ODT)
    222
Organic mass 52, 56
Orifice diameter 68, 76

P

Plasma enhanced chemical vapor
    deposition (PECVD) 117
Plasma-enhanced chemical vapor
    deposition (PECVD) 115
Polyethylene terephthalate (PET)
    144, 146, 152

R

Red luminescence (RL) 211

S

Scanning electron microscopy
    (SEM) 153, 174
Strainer diameter 54

T

Thermionic vacuum arc (TVA)
    116, 117, 137
Thermo gravimetric (TG) 144
Thin Layer Chromatography (TLC)
    221
Tonnes per hour (TPH) 40
Transmission electron micros-
    copy (TEM) 63
Transverse electromagnetic mode
    (TEM) 84

U

User-defined function (UDF) 4

V

Vacuum evaporation method 50
Vapor-liquid-solid (VLS) 192,
    193, 200

W

Waste PET lightweight aggregates

(WPLA) 146
Waste production 146

Y

Yellow luminescence (YL) 211